Reaction Kinetics

Ernő Keszei

Reaction Kinetics

An Introduction

Ernő Keszei
Department of Physical Chemistry
Eötvös Loránd University
Budapest, Hungary

ISBN 978-3-030-68573-7   ISBN 978-3-030-68574-4 (eBook)
https://doi.org/10.1007/978-3-030-68574-4

© The Editor(s) (if applicable) and The Author(s), under exclusive license to Springer Nature Switzerland AG 2021

This work is subject to copyright. All rights are solely and exclusively licensed by the Publisher, whether the whole or part of the material is concerned, specifically the rights of translation, reprinting, reuse of illustrations, recitation, broadcasting, reproduction on microfilms or in any other physical way, and transmission or information storage and retrieval, electronic adaptation, computer software, or by similar or dissimilar methodology now known or hereafter developed.

The use of general descriptive names, registered names, trademarks, service marks, etc. in this publication does not imply, even in the absence of a specific statement, that such names are exempt from the relevant protective laws and regulations and therefore free for general use.

The publisher, the authors, and the editors are safe to assume that the advice and information in this book are believed to be true and accurate at the date of publication. Neither the publisher nor the authors or the editors give a warranty, expressed or implied, with respect to the material contained herein or for any errors or omissions that may have been made. The publisher remains neutral with regard to jurisdictional claims in published maps and institutional affiliations.

This Springer imprint is published by the registered company Springer Nature Switzerland AG
The registered company address is: Gewerbestrasse 11, 6330 Cham, Switzerland

Preface

Learning (and teaching) the theoretical basis of chemical kinetics is not an easy task. Following the diversification of master's programmes, shorter undergraduate programmes – typically three years in most European countries – do not provide sufficient knowledge in physical chemistry courses to allow students to gain a deeper insight of chemical kinetics. A textbook for beginners, which would contain enough detail but avoid outdated theories and methods, is not really available. Though there are several excellent textbooks on more advanced topics in reaction kinetics, they typically presuppose quite a good knowledge of theoretical basics, so that their understanding poses problems for the average graduate student.

A modern textbook should accommodate the needs of such students. In addition, at the end of the twentieth and beginning of the twenty-first century, developments in chemical kinetics, both experimental and theoretical, mean that it is no longer necessary to emphasize the customary approach based on the "order" of reaction, nor to oversimplify emerging systems of ordinary differential equations. Traditional, not fully justifiable simplification of the kinetic equations and associated "linearization" of functions are unnecessary, as there are plenty of numerical integrators and other nonlinear numerical methods that are readily available. Such traditional approaches are helpful for the students insofar as they help in understanding old research papers and results.

Similarly, "formal kinetics" does not need to be introduced in its old-fashioned "ad hoc" version; since chemistry nowadays is a truly molecular science, having solid quantum mechanical and statistical thermodynamic foundations, chemical kinetics should also be based on this knowledge. Thus, the introduction to the subject should emphasize transition state theory, which is the most widely used and still powerful representation of the kinetic aspects of elementary chemical reactions.

According to these considerations, the goal of this text is to introduce the student to the science of temporal evolution of chemical reactions by a molecular-statistical interpretation, and to refer to this picture whenever possible. The author hopes that students reading this text will find it easy to understand and – at the same time – will acquire a genuine knowledge of reaction kinetics that is readily applicable in practice.

To achieve these goals, the book is organized into seven chapters. After the introductory Chap. 1, two theories of current interest – collision theory and transition state theory – are described in enough detail in Chap. 2 for calculating rate coefficients as well as providing background for a better understanding of the later chapters. Chapter 3 describes traditional "formal kinetics" that can be applied to elementary reactions as well as to individual reaction steps in composite reactions. Details of solving simple mass action rate equations to provide explicit concentration vs time functions are also discussed. To find kinetic parameters, the application of nonlinear models is emphasized, but traditional linearization methods are also explained to help the reader understand old research papers. Chapter 4 deals with the kinetics of composite reactions. Examples of basic types of reaction mechanisms are treated in detail, along with solutions of the emerging rate equations. Simplification of kinetic models using different approximations is also part of this chapter. Along with some methods to obtain analytical solutions, numerical solution and related parameter estimation is also treated in enough detail to help the reader use kinetic software packages which can solve complicated mechanisms and estimate their relevant parameters. Chapter 5 is about activation of reactants to initiate reactions, with particular attention given to gas phase thermal activation of unimolecular processes. Chapter 6 deals with catalysis at an elementary level, discussing heterogeneous reactions on solid surfaces and enzyme catalysis in somewhat more detail. A brief discussion of autocatalytic processes and oscillating reactions is also included. Chapter 7 is the last one and describes experimental methods in reaction kinetics; it includes most methods and reactors used nowadays within the timescale from a few days down to a few femtoseconds.

Forty-five colour figures aid understanding of the content. There are end-of-chapter solved problems that either illustrate the application of the content of the chapter or explain interesting further details in the topic itself. Further reading is provided at the end of each chapter. Books and papers listed refer to texts that either treat the topic of the chapter in more detail or with a different approach, or contain relevant information concerning the solved problems.

The material of the book is enough for a substantial undergraduate reaction kinetics course, or for part of a detailed physical chemistry curriculum. However, some parts or even chapters can be omitted in courses for non-chemistry-major students who need only a limited knowledge of chemical kinetics, but access to the creative use of kinetic principles.

Vladimir Nabokov, the Russian-English-American novelist wrote the following about his adopted language: "I had to abandon my natural idiom, my untremmeled, rich and infinitely docile Russian tongue for a second-rate brand of English, devoid of any of those apparatuses – the baffling mirror, the black velvet backdrop, the implied associations and traditions – which the native illusionist, frac-tails flying, can magically use to transcend the heritage in his own way." My English is certainly at least as second-rate as that of Nabokov; however, I hope that it would not disturb too much native English speakers, and would not make understanding concepts explained in this book difficult for students.

This material is a result of 10 years of teaching chemical kinetics as part of a new undergraduate physical chemistry curriculum. During this time, it has been improved through my experience gained from teaching students. A number of them contributed to the improvement of the curriculum and the explanatory part. I am also indebted to colleagues who helped to enhance the quality of the text. I would like to mention five of them here: Gábor Czakó, who performed good-quality quantum mechanical calculations to provide a reliable PES of the collinear H-exchange reaction; Tibor Nagy, who performed semiclassical calculations to obtain relevant molecular parameters for the iodine molecule consisting of ^{127}I and ^{125}I atoms; Soma Vesztergom, for a critical reading of the end-of-chapter problems; Vilmos Gáspár, for his valuable advice to keep the section on autocatalysis and related phenomena concise and easy to understand; and Éva Valkó, who helped to find errors in published data on the kinetics of hydrogen bromide formation.

Since the time I first presented the text to Springer Nature, I have experienced valuable help of Dr. Angeliki Athanasopoulou, editor in chemistry. I would like to thank her for her continuous care for this project, which helped me to produce the manuscript within the deadline. I am also grateful to Ms. Arul Vani Parttibane for her understanding and successful efforts to produce text and especially formulae in the correct shape.

Budapest, Hungary
October 2020

Ernő Keszei

Contents

1	**Introduction**	1
	Further Reading	4
2	**Theories of Elementary Reactions**	5
	2.1 Collision Theory	7
	2.2 Transition State Theory	10
	2.2.1 Potential Energy Surfaces in Reactive Systems	10
	2.2.2 Statistical Formulation of the Equilibrium Constant	14
	2.2.3 Quasi-Equilibrium Transition State Theory	19
	2.2.4 Dynamical Treatment of the Transition State Theory	26
	2.3 Dependence of the Rate Coefficient on Temperature and Pressure	29
	Further Reading	37
3	**Formal Kinetic Description of Simple Reactions**	39
	3.1 Solution of Rate Equations of Integer-Order Reactions	41
	3.1.1 Zero-Order Reactions	47
	3.1.2 First-Order Reactions	48
	3.1.3 Second-Order Reactions	50
	3.1.4 Third-Order Reactions	57
	3.2 Generalisation and Extension of the Order of Reaction; Pseudo-Order	61
	Further Reading	70
4	**Kinetics of Composite Reactions**	71
	4.1 Coupling of Elementary Reactions	72
	4.2 Parallel Reactions	74
	4.3 Consecutive Reactions	77
	4.4 Reversible Reactions	80
	4.5 Solving Rate Equations of Mechanisms Comprising First-Order Reactions Only	84
	4.6 Quasi-Steady-State Approximation	88

	4.7	Fast Pre-equilibrium Approximation	90
	4.8	Rate-Determining Steps	93
	4.9	Numerical Algorithms to Solve Differential Equations; Integrators for Reaction Kinetics	95
	4.10	Chain Reactions	99
		4.10.1 Unbranched Chain Reactions	99
		4.10.2 Branched Chain Reactions and Explosions	105
	Further Reading		116

5 Activation Processes and Unimolecular Gas Phase Reactions ... 117
    5.1 Molecular Interpretation of Activation ... 117
    5.2 Theories of Unimolecular Gas Reactions ... 119
        5.2.1 Lindemann Mechanism ... 120
        5.2.2 Lindemann-Hinshelwood Mechanism ... 122
        5.2.3 RRK Theory ... 125
        5.2.4 RRKM Theory ... 127
    Further Reading ... 131

6 Catalysts and Catalytic Reactions ... 133
    6.1 Heterogeneous Catalysis ... 136
    6.2 Enzyme Catalysis ... 142
    6.3 Autocatalysis, Autoinhibition and Nonlinear Chemical Processes ... 147
        6.3.1 A Simple Autocatalytic Reaction ... 148
        6.3.2 Oscillating Reactions ... 151
    Further Reading ... 159

7 Experimental Methods in Reaction Kinetics ... 161
    7.1 Methods to Determine Concentration ... 163
    7.2 Initiating Reactions at Different Timescales ... 167
    7.3 Experimental Techniques and Devices ... 168
        7.3.1 Classical Techniques ... 168
        7.3.2 Flow Methods ... 170
        7.3.3 Relaxation Methods ... 174
        7.3.4 Flash Photolysis and Laser Photolysis ... 176
        7.3.5 Kinetic Measurements at Femtosecond Timescale ... 178
    Further Reading ... 181

Index ... 183

Chapter 1
Introduction

Reaction kinetics is the branch of physical chemistry concerned with the temporal evolution of chemical reactions. It provides explanation as to why some reactions do not take place, although their products would be much more stable thermodynamically than the reactants are; it accounts for the fact that some reactions go very fast while others rather slowly. It also provides several useful tools to calculate the actual rate of reactions under a wide variety of conditions, including their dependence on temperature, pressure, the solvent applied etc. An important topic also treated by reaction kinetics is the initiation of reactions – usually called *activation* – using different effects (e.g. light, heat, radioactivity, ultrasound), that is, the acquisition of enough energy by the molecules to start a reaction.

The most important goal of kinetic research is the identification and characterisation of elementary molecular events that make possible the transformation of reactants into products. At the time of the beginning of reaction kinetic studies – in the second half of eighteenth century when mechanical models dominated natural sciences – the scheme of these simple molecular events that constitute the overall reaction has been called the *mechanism* of the reaction, which term survived and is still widely used. This means that the *elementary* steps of the reaction – the simplest events when typically two molecules directly encounter – are identified, and the complex reaction mechanism is constructed combining these elementary steps. Most real-life chemical reactions (called *composite reactions*) comprise quite a number of such elementary steps; the number of them, for example, in gas reactions, can be as much as a few hundreds. The overall rate of these reactions can successfully be calculated over a wide range of conditions if we can calculate the rate of all the constituent elementary reactions under the given conditions, and also know the way of their connection within the reaction mechanism. This is the reason why the theory of elementary reactions is of paramount importance in chemical kinetics.

However, to explore the precise mechanism of composite reactions is not an easy task. To be able to construct a reliable mechanism, we have to identify all the components that take part in the reaction, even if they are short-lived and have rather low concentration. We also have to keep track of the temporal evolution of possibly all of these components, although this is not always possible. In many reactions, there are so called *intermediates* (substances formed from the *reactants* and readily removed in subsequent reactions leading to the *products*) that are rather short-lived and present only in very low concentrations; thus, they might easily remain unnoticed for the experimentalist. After having traced the temporal evolution of as many components of the reacting mixture as possible, the task of the kineticist is to construct a suitable mechanism that can explain all the concentration changes as a function of time. Consequently, theories of composite reactions also constitute an important topic of reaction kinetics.

In order to successfully model reactions the way explained above, we need an unambiguous definition of the rate of change of the amount of components taking part in the reaction. There exists an International Union of Pure and Applied Chemistry (IUPAC) recommendation for the definition of the *rate of reaction*. For the sake of this definition, we shall write stoichiometric equations in a special form, so that the equation is set equal to zero. The advantage of this form is that reactants and products (species on the left-hand side and on the right-hand side in the more common equation) can be treated the same way, simplifying the mathematical treatment. The general stoichiometric equation of this form can be written as:

$$\sum_{i=1}^{R} \nu_i A_i = 0 \qquad (1.1)$$

The symbol A_i denotes the stoichiometric formula of the species i, and ν_i (lower case Greek 'nu') is the *stoichiometric number* of this species. The index i runs over all the *reacting species* whose number is R. (Components that do not react – e.g. an inert solvent – should have a zero stoichiometric number, thus it is superfluous to include them in the sum.) As an example, let us write one of the possible equations of the formation of water:

$$-1\ H_2 - \tfrac{1}{2}\ O_2 + 1\ H_2O = 0 \qquad (1.2)$$

In this equation, $A_1 = H_2$, $A_2 = O_2$, $A_3 = H_2O$, $\nu_1 = -1$, $\nu_2 = -\tfrac{1}{2}$ and $\nu_3 = 1$. However, we are too much used to write stoichiometric equations in the usual left side–right side form. Thus, we usually write the equation itself in the traditional form,

$$H_2 + \tfrac{1}{2}\ O_2 = H_2O, \qquad (1.3)$$

but we consider the stoichiometric number ν_i of the reactants (left side) to be *negative*, while those of the products (right side) to be *positive* – as if all the terms of the equation were arranged to the right side. In the rest of this book, we always interpret stoichiometric equations this way.

The rate of a particular reaction should be defined naturally in such a way that it should be independent from the choice of the component taking part in the reaction whose amount would be used to account for the temporal evolution. To formulate such a definition, let us introduce the *extent of reaction* ξ (also used in thermodynamics); but for use in the definition of the rate of reaction, it is sufficient to specify its change as:

$$d\xi = \frac{dn_i}{\nu_i}. \tag{1.4}$$

The definition of the *rate of the reaction*, which is written in the above explained general form (1.1), can then be written as:

$$r = \frac{d\xi}{dt} = \frac{1}{\nu_i}\frac{dn_i}{dt} \tag{1.5}$$

According to this, the SI unit of the rate of reaction is mol (stoichiometric equation)/s. In the chemical praxis, concentrations are much more convenient to measure than amounts of substances. Taking this into account, let us calculate from the above definition the rate of change of molar concentration of a species. Using the definition of the molar concentration as $c_i = n_i/V$, we get

$$\frac{1}{\nu_i}\frac{d(c_i V)}{dt} = \frac{1}{\nu_i}\left(V\frac{dc_i}{dt} + c_i\frac{dV}{dt}\right). \tag{1.6}$$

This equation reveals that the rate of change of the molar concentration dc_i/dt also depends on the rate of change of volume. If the volume does not change during the reaction, the rate of reaction can be obtained by multiplying dc_i/dt by the volume of the reacting system, and dividing it by the stoichiometric number. Accordingly, we can state that *in case of reactions at constant volume*, the change of molar concentration is identical to the stoichiometric number times the *rate of reaction divided by the volume*:

$$\frac{dc_i}{dt} = \frac{1}{V}\frac{dn_i}{dt} = \frac{\nu_i}{V}\frac{d\xi}{dt} \tag{1.7}$$

Within chemical kinetic context, the quantity dc_i/dt is usually called simply as the rate of reaction. However, we should be aware that this quantity is proportional only to the rate of reaction in case of constant volume reactions, when the proportionality constant is the volume V divided by the stoichiometric number ν_i. In condensed phase reactions (e.g. in solutions) this is typically true to a good approximation, thus the proportionality can be assumed. Further on in this text, we also consider dc_i/dt as the rate of the reaction. However, in gas phase reactions, if there is a change in number of moles during the reaction, this proportionality does not hold and the change of volume should also be taken into account. It is also worth mentioning that the latest recommendations of IUPAC suggest to use the term 'rate of conversion' for $d\xi/dt$ and the 'rate of reaction' for $\frac{1}{\nu_i}\frac{dc_i}{dt}$. Anyway, naming of the terms does not change their relation as explained above.

As we have seen, the definition of the reaction rate (in both versions) refers to a particular stoichiometric equation as it contains the corresponding stoichiometric numbers. It is worth mentioning that even this definition is only valid if during the (composite) reaction, there will not be any accumulation of an intermediate, and no other 'by-products' are formed besides those included in the stoichiometric equation. (This is the case for elementary reactions where this definition can always be used.) Thus, in case of composite reactions, it is more convenient to refer to the change of the amount (or concentration) of a *component*, or equivalently to the *rate of consumption* for reactants and the *rate of appearance* for products.

Further Reading

1. Pilling MJ, Seakins PW (1995) Reaction kinetics. Oxford University Press, Oxford
2. de Paula J, Atkins PW (2014) Physical chemistry. 10th edn. Oxford University Press, Oxford
3. Silbey LJ, Alberty RA, Moungi GB (2004) Physical chemistry. 4th edn. Wiley, New York
4. Steinfeld JI, Francisco JS, Hase WL (1998) Chemical kinetics and dynamics. 2nd edn. Prentice Hall, Englewood Cliffs
5. Compendium of Chemical Terminology/Gold Book Version 2.3.3 (2014), International Union of Pure and Applied Chemistry

Chapter 2
Theories of Elementary Reactions

Thermodynamics tells us that a chemical reaction taking place in a closed, rigid and constant temperature container sooner or later reaches equilibrium where the free energy F of the reacting mixture is minimal,[1] as a function of the concentrations. However, we also know that – quite often – this does not happen. Let us imagine such a container filled with hydrogen and oxygen at room temperature and atmospheric pressure; there will not be any reaction between the two components and the system would stay in this metastable state for any length of time provided that conditions would not change. Thinking in terms of thermodynamics, metastable equilibria in (composite) systems can usually be maintained easily by separating the subsystems with a suitable wall. The constraint exerted by this wall can maintain the metastable equilibrium that – without the presence of the wall – would be transformed into a stable equilibrium.

Returning to the example of hydrogen and oxygen, let us imagine two compartments of a closed, rigid, thermostated system with a compartment containing hydrogen twice as large as the other one containing oxygen, both having the same pressure. If the subsystems are also closed, that is, the interior wall does not allow for material transport, the metastable state without even mixing the two components would be maintained. In this state, both hydrogen molecules and oxygen molecules move randomly within the volume occupied, having frequent collisions with each other. Let us remove the separating wall with a suitable method (e.g. breaking the wall with an external actuator), and wait for a while. What happens after the removal of the wall? Molecules do continue the same random movement as before, and – as a consequence of this random movement – they will mix completely to form a

---

[1]Here, F is the constant temperature–constant volume potential (*free energy* or *Helmholtz potential*), which is minimal in a constant temperature and constant volume system at thermodynamic equilibrium. A frequent alternative notation is A; after the German name '(nützliche) *A*rbeit' – in English: (useful) work. Not to mix with the Gibbs potential denoted by G, which is the constant temperature–constant pressure potential, and is minimal at constant temperature and pressure in systems at thermodynamic equilibrium.

homogeneous mixture that is stoichiometric from the point of view of water formation, regarding the hydrogen:oxygen ratio.

It is worth to draw some conclusions at this point. There are some textbooks – not considering the real nature of mixing at molecular level – which state that hydrogen and oxygen *have a tendency* to mix after the removal of the separating wall. Well, a gas consists of *molecules*, thus any 'tendency' can only be attributed to molecules of the gas. However, gas molecules are not bothered by this change; they have a fast random motion with frequent collisions before, as well as after the removal of the separating wall. In other words, they *do not have any tendency*. *This is exactly the reason* for a homogeneous mixing. Should they have any other 'tendency' in addition to their random movement, they would *not* mix homogeneously.

There are other (educational) resources emphasising that a *chemical equilibrium is dynamic*. This statement is based on the fact that a chemical reaction in equilibrium proceeds with equal rate to both directions. We can easily realise that this statement is valid for *any equilibrium* where molecules take part. In the previous example, molecules move randomly all over the accessible volume in both compartments of the container. In equilibrium, the same number of them will wander over to one part of the container on average, as the number that wanders to the opposite direction; which guarantees the maintenance of a homogeneous distribution. Accordingly, the (metastable) equilibrium is *dynamic*; molecules are constantly running back and forth. Exactly the same happens when the separating wall is removed. The vigorous random motion of molecules leads quite quickly to a homogeneous distribution within the enlarged volume, and thus to a perfect mixing. This new equilibrium is also dynamic; on average, the same number of hydrogen molecules will move away from the surroundings of oxygen molecules, as those that move towards them in the opposite direction. Thus, we can conclude that *every macroscopic equilibrium is microscopically dynamic*. (As a matter of fact, we should also note that in microscopic volume elements, there are of course *fluctuations*, but they cancel within greater [macroscopic] volumes.)

Let us return to the stoichiometric hydrogen–oxygen mixture. According to thermodynamic calculations, the constant volume–constant temperature mixture is largely metastable, as the free energy of formation of water (at room temperature) is -3006 kJ/mol. This would make possible an important decrease in free energy on water formation with respect to hydrogen and oxygen, as their free energy of mixing is only a meagre -158 kJ/mol. Despite this large driving force, the reaction does not take place. Remembering the case with mixing after the removal of the separating wall we could think that there is some constraint that prohibits the process to the chemical equilibrium, that is, water formation. Obviously, there is no macroscopic 'wall' that hinders the reaction in the homogeneous mixture; we should rather look for a 'molecular-level' hindrance. Furthermore, we also know from chemistry that placing, for example, porous platinum into the mixture, water formation would quickly happen. This means that the 'molecular-level' wall can also be removed, though – as we shall see later – catalysts typically do not 'remove' the wall, rather provide a way for molecules to bypass it.

This chapter deals with the details of the 'molecular wall', which prohibits chemical reactions, and also with the means molecules have at their disposal to 'climb' that wall.

2.1 Collision Theory

There are several theoretical approaches to describe chemical reactions at a molecular level. One of the simplest is the kinetic theory of gases that provides a way to calculate the collision frequency of molecules. In addition, we can also calculate the portion of molecules that has enough energy to overcome the resistance originating in the stability of the reactants, thus promoting the formation of reaction products.

Let us consider the general gas reaction:

$$A + B \rightarrow \text{products} \tag{2.1}$$

To calculate the reaction rate, we use the following model. Molecules are considered as hard elastic spheres. The motion of a molecule A in the direction of a fixed molecule B is described in a coordinate system with its origin at the centre of mass of molecule B (see Fig. 2.1). In this coordinate system, molecule A approaches molecule B with a relative velocity $v = v_A - v_B$. Let us denote the radius of molecule A by r_A, that of molecule B by r_B, and the sum of the two by $d = r_A + r_B$. Distance b – called the *impact parameter* – is the closest perpendicular distance of the centres of masses of molecules A and B along the trajectory of molecule A. If $b > d$, A passes by B without collision, but if $b < d$, A collides with B and gets deflected. In other words, if the centre of mass of B is within a cylinder of radius d around the trajectory of the centre of mass of A, collision happens, but if it is without, no collision

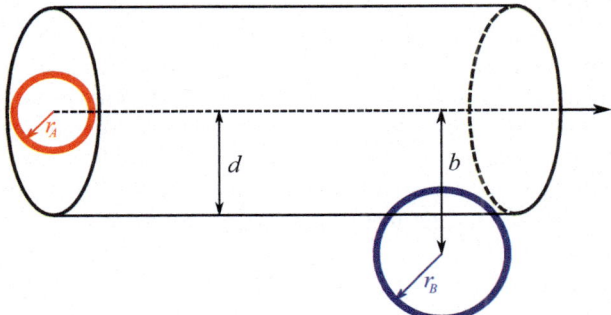

Fig. 2.1 Notation for the description of collisions between molecule A (red) and molecule B (blue). The dashed line indicates the trajectory of molecule A approaching the stationary molecule B. Collision occurs if the centre of mass of B is within the collision volume indicated by the cylinder of radius d. In the figure, $b > d$; thus, there is no collision

happens. (A usual way of expressing this in a quantitative way is that the *total collision cross-section* or *effective target area* of B presented to A is $S = d^2\pi$.)

The travelling molecule A thus sweeps $d^2\pi v$ 'collision volume' within unit time. The number of its collisions with molecules B depends on the *number density* N_B of molecules B. If there are N_B molecules B in one volume unit gas, then one molecule A collides with $N_B d^2 \pi v$ molecule B within unit time.[2] This is valid for each molecule A; thus, the overall number of collisions between A and B per unit volume within unit time can be given as $Z' = N_A N_B d^2 \pi v$, where N_A is the number density of molecules A.

The expectation value $M(v)$ of the velocity of molecules can be calculated using the Maxwell distribution for the absolute value of their velocity. If we perform this calculation for the relative velocity of two molecules, we should use the *reduced mass*,[3] which can be given with the masses of the individual molecules as

$$\mu = \frac{m_A m_B}{m_A + m_B}. \tag{2.2}$$

Substituting this into the expression of the expectation value we get:

$$M(v) = \sqrt{\frac{8 k_B T}{\pi \mu}} \tag{2.3}$$

where T is temperature and k_B the Boltzmann constant. Writing this mean velocity into the expression of Z', we get the result

$$Z' = N_A N_B d^2 \pi \sqrt{\frac{8 k_B T}{\pi \mu}}. \tag{2.4}$$

This is the total number of collisions, irrespective of whether there will be a reaction or not between the colliding molecules. Colliding molecules whose collision energy along the direction of collision is below a threshold E_a cannot surmount the *energy barrier* necessary for the reaction; thus, they undergo an *elastic collision*. Molecules whose collision energy along the direction of collision exceeds this value will undergo an *inelastic collision* leading to reaction. (These are called *reactive collisions*.) Accordingly, to calculate the reaction rate, we have to calculate the portion of collisions leading to reaction. For this purpose, we use the Boltzmann distribution describing the probability density of kinetic energy. According to this, the proportion of collisions having greater than E_a energy can be given as

---

[2]Evidently, the direction of the moving molecule changes with each collision, but the collision volume will be a cylinder of radius d further on as well; thus, this would not change the validity of the formula.

[3]The definition of the reduced mass is $\frac{1}{\mu} = \frac{1}{m_A} + \frac{1}{m_B}$, which is equivalent to Eq. (2.2).

2.1 Collision Theory

$$\frac{Z'_{E_a}}{Z'} = e^{-\frac{E_a}{RT}}, \qquad (2.5)$$

where Z'_{E_a} is the number of collisons with higher energy than E_a, while Z' is the total number of collisions.

Multiplying the total number of collisions Z' by this ratio, we get R, the number of reactive collisions per unit time, that is, the rate of reaction. After some rearrangement, this rate can be written in the following form:

$$R = d^2 \pi \sqrt{\frac{8 k_B T}{\pi \mu}} e^{-\frac{E_a}{RT}} N_A N_B \qquad (2.6)$$

We can see that the first four factors do not depend on the concentration (i.e. on the number densities of A and B); thus the product of these factors can be considered as a coefficient independent of concentration what we call *rate coefficient* (also called as *rate constant*):

$$k = d^2 \pi \sqrt{\frac{8 k_B T}{\pi \mu}} e^{-\frac{E_a}{RT}} \qquad (2.7)$$

Using this notation, the rate of the reaction can be written as $R = k N_A N_B$. We can sum up this result the following way. The rate of reaction (2.1) is proportional to the product of the concentration of the reactants, containing the concentration-independent proportionality factor k. This latter can be given by Eq. (2.7), which is the expression of the *rate coefficient according to collision theory*. It is worth noting that the literature on gas reactions typically uses the unit molecules/(cm^3 s) for the reaction rate – as a consequence of the above discussed calculations. The coefficient $d^2 \pi \sqrt{\frac{8 k_B T}{\pi \mu}}$ of the exponential function – originating from the collision number – is usually denoted by Z and called the *collision frequency factor*. Comparing its expression to Eq. (2.6), we can alternatively express it as $Z = Z'/(N_A N_B)$. Using this, the rate coefficient can be written as the product of the collision frequency factor and the Boltzmann factor:

$$k = Z\, e^{-\frac{E_a}{RT}} \qquad (2\,8)$$

Rate coefficients and related reaction rates calculated from the collision theory expression naturally inherit all approximations used in the above derivation. The most simplifying approximation is to treat molecules as small elastic spheres and using classical mechanical laws of elastic collisions. Molecules are of course more complicated entities. Another, rather simplifying approximation is to simply take into account collisions above a threshold of kinetic energy to lead to reaction, which does not take into account the structure of the molecules. According to this, we expect that the rate coefficients and reaction rates calculated this way are best valid for reactions between *monatomic reactants* – which is in fact supported by experimental evidence.

2.2 Transition State Theory

Unlike collision theory, transition state theory (TST) takes into account the internal structure of molecules, in both non-reactive and reactive collisions. Its basic idea is that the fate of colliding molecules is determined by the actual location of all atoms within, along with their interactions. Whether the outcome is a reaction, as well as the rate of the eventual reaction itself, depends on these interactions. There are several versions of the transition state theory; the historically first and most simple being the so-called *quasi-equilibrium* description, which we shall explore next. Before going into details of the theory, we shall refresh relevant statistical thermodynamic relations that make it possible to describe systems of constant temperature and constant volume, and also constant temperature and constant pressure systems. However, as TST is also based on potential energy surfaces, let us discuss them first.

2.2.1 *Potential Energy Surfaces in Reactive Systems*

Transition state theory (TST) is based on the hypothesis that interactions of colliding molecules can be described in terms of the interactions of all the atoms contained in the reactant molecules. However, this interaction is not described in terms of time but the spatial arrangement of the interacting atoms that we call *configuration*. The relevant quantity that would govern the fate of the interacting atoms is the *potential energy* as a function of this configuration. A convenient visualisation method is to plot the potential energy as a function of the spatial coordinates of the atoms. This (multidimensional) surface is called the *potential energy surface*, which we shall denote by the acronym *PES*.

The simplest PES is a two-dimensional curve, the *potential energy curve of a diatomic molecule*. This curve can be seen on Fig. 2.1 for the case of a ground-state hydrogen molecule. Negative potential energy values indicate attractive force, positive values repulsive force between the atoms. It can be seen from the figure that the minimum of the potential energy curve is at the equilibrium distance of 76.2 pm. Placing the hydrogen atoms closer to each other, attraction is decreasing radically with decreasing distance and it changes into repulsion within small distances. The curve then rises very steeply, indicating strong repulsion. Placing the two atoms further from each other than the equilibrium distance, their attractive interaction diminishes with increasing distance. At large enough distances, their interaction becomes zero and the molecule can be considered as dissociated.

Chemical reactions can be interpreted in the PES diagrams as changes in the spatial arrangement (*configuration*) of the atoms; that is, a movement of the

2.2 Transition State Theory

configuration point on the surface from the arrangement of *reactants* until that of the *products*. Stable molecules can always be found at the bottom of 'valleys' characterised by a minimum in potential energy; thus, the reaction path is a trajectory from the reactant valley to the product valley. Accordingly, the rate of reaction can be determined by calculating the speed at which the atoms taking part in the reaction arrive from the reactant valley (on the surface of the PES) to the product valley. To calculate this rate, we can use several methods. We shall deal first with the simplest one, which is based on the presupposition that the thermodynamic equilibrium is always maintained between the reactants and a specific configuration we call *transition state*. This method is called the *quasi-equilibrium* description.

Coming back to the PES; in general, it is a multidimensional surface that shows the potential resulting from the interaction of all the atoms in the reacting molecules as a function of the configuration of the atomic nuclei, for all the relevant configurations that could occur during the reaction. To illustrate this surface – for practical reasons – we should choose a simple case. One of the first surfaces calculated has been the hydrogen exchange reaction $H_2 + H$, with some further simplification, namely in its *collinear* form. This means that the reaction

$$H_2 + H \rightarrow H + H_2 \tag{2.9}$$

is restricted to take place in a way when the three hydrogen atoms involved in the reaction are always located along a line; that is, the reaction proceeds via a 'head-on' collision only. This restriction enables us to uniquely specify the configuration by two interatomic distances only, and, as a result, we can visualise the emerging PES in a tridimensional diagram. (Without the collinear restriction, we would have to use a third coordinate, for example, an angle characterising the orientation of the H_2 molecule with respect to the H atom.) (Fig. 2.2)

Let us choose the distances of the two terminal hydrogen atoms from the medial one as the two coordinates to describe the configuration. Identifying the hydrogen atoms by the letters A, B and C, one distance is between the nuclei of A and B, the other one between those of B and C. (Let us call them simply as distance A–B and distance B–C.) The PES diagram constructed this way shows the potential energy as a function of these two distances. The reaction itself can be represented by the simple scheme seen on Fig. 2.3.

Prior to the reaction, the (non-rotating) molecule AB and the lone hydrogen atom C approach each other along the direction of the bond between the atoms A and B. When the actual collision occurs, the newly formed (non-rotating) molecule BC and the atom A recoil, according to the laws of the conservation of energy and momentum.

We can track this reaction on the PES diagram in terms of the two distances chosen. In the region of configurations characteristic of the reactant molecule AB, the section of the PES along the direction A–B at constant (large) distances B–C is expected to show the same shape as that of an equilibrium ground-state potential energy curve of a hydrogen molecule, as seen in Fig. 2.2. We also expect a similar shape of the section along the direction B–C at large distances A–B. The PES shown

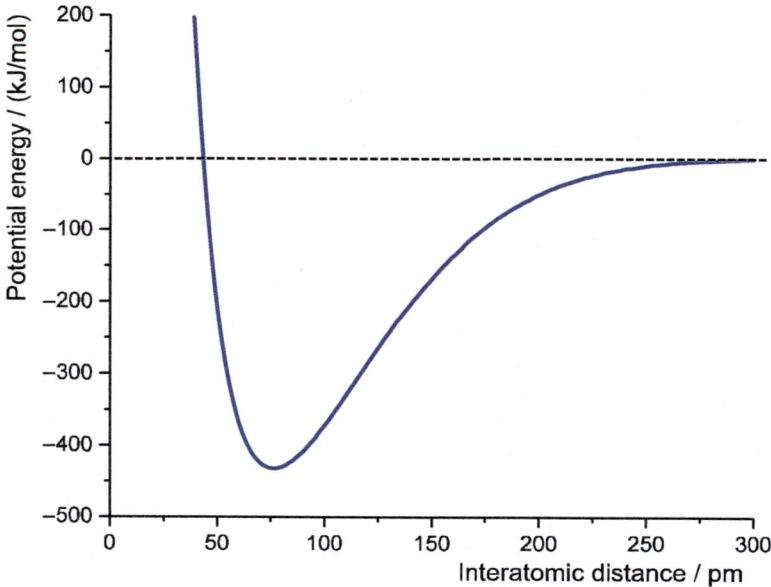

Fig. 2.2 Potential energy of a ground-state hydrogen molecule as a function of interatomic distance

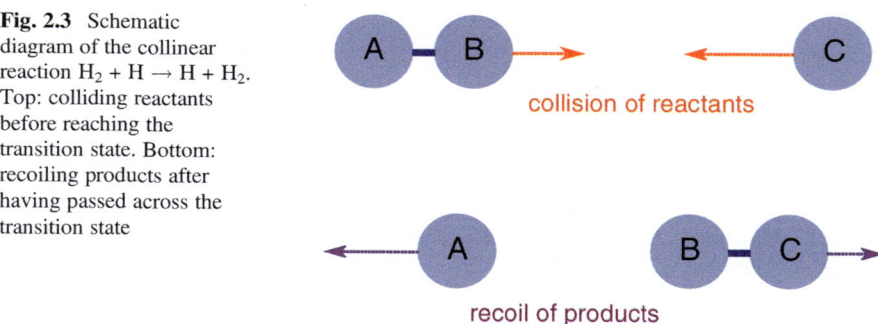

Fig. 2.3 Schematic diagram of the collinear reaction $H_2 + H \rightarrow H + H_2$. Top: colliding reactants before reaching the transition state. Bottom: recoiling products after having passed across the transition state

as a perspective on Fig. 2.4 has been constructed based on an efficient quantum mechanical calculation.[4] At small internuclear distances, a low valley is clearly seen, which extends from the reactant valley at small A–B and large B–C distances (representing the reactant AB molecule and a distant C atom) to the product valley at large A–B and small B–C distances (representing the product BC molecule and a distant A atom).

---

[4]The method of calculations has been the full configuration interaction using augmented correlation-consistent polarised basis; abbreviated as full-CI/aug-cc-pVDZ. (Calculations have been performed by Gábor Czakó. All further data concerning this reaction are from these calculations.)

2.2 Transition State Theory

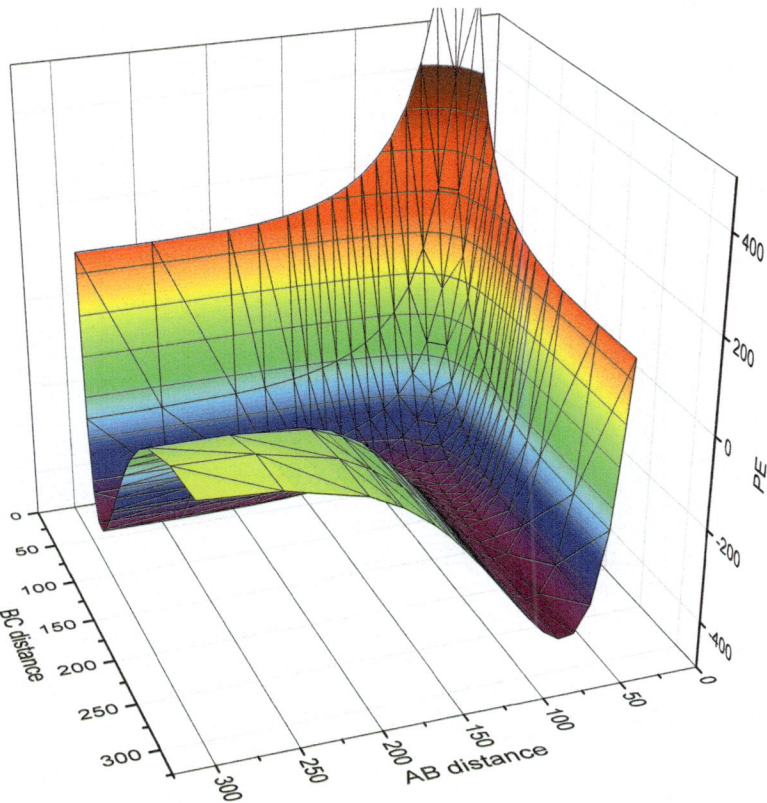

Fig. 2.4 Potential energy surface of the collinear reaction $H_2 + H \rightarrow H + H_2$: potential energy of three collinearly arranged hydrogen atoms (A, B and C) as a function of interatomic distances. Sections at 300 pm A–B or B–C distance are practically identical to the diatomic potential energy curve of a stable hydrogen molecule

On the perspective seen in the figure, we cannot have a proper look inside the bent valley. However, similarly to topological maps, we can project the tridimensional PES into a contour map, which can be completely shown on the printed page or a flat screen. This can be done by projecting lines connecting equipotential loci onto a horizontal plain in the tridimensional diagram. This contour diagram contains all information about the fully tridimensional surface. As the bottom of the bent valley in this case is quite shallow, the diagram shown in Fig. 2.5 is constructed in a way that contour lines at the bottom of the valley are closer to each other (red dashed lines; at 5 kJ/mol level differences) than those further away from the bottom of the valley (blue continuous lines; at 50 kJ/mol level differences).

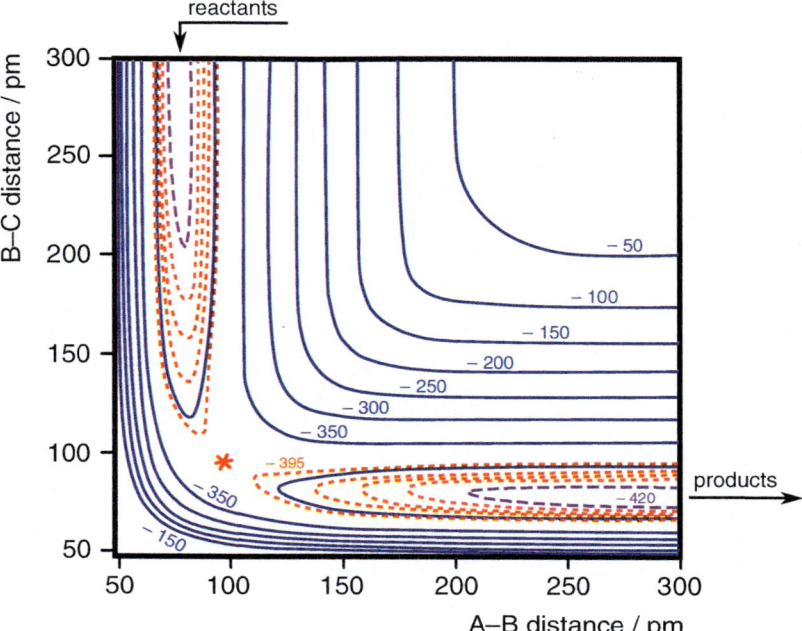

Fig. 2.5 Contour map of the potential energy surface of the collinear reaction $H_2 + H \rightarrow H + H_2$. The distance in level of the continuous blue lines is 50 kJ/mol. Red dashed lines show elevations ten times closer, at 5 kJ/mol distances

2.2.2 Statistical Formulation of the Equilibrium Constant

Within the context of the quasi-equilibrium transition state theory, it is presupposed that the reaction leading to the transition state is in thermodynamic equilibrium. We shall describe this equilibrium using molecular properties of the reacting species, based on the results of statistical thermodynamics. Before actually applying these results for the case of transition state formation, let us briefly recall relevant formulae to use for the calculation of the equilibrium constant of chemical reactions at constant temperature and pressure. For these conditions we shall need partition functions in canonical ensembles.

A canonical ensemble can be characterised by its temperature T, volume V and composition. (This latter will be given as the *number* of molecules N, instead of the amount of substances.) Accordingly, a canonical ensemble is such a collection of virtual states of a large number of particles, whose volume, temperature and number of particles is fixed. (It represents a thermodynamic system of fixed volume and fixed composition, which is in thermal equilibrium with a thermostat at fixed temperature.) Alternatively, the canonical ensemble can also be called an N, V, T ensemble. On this ensemble, the *free energy function*

2.2 Transition State Theory

$$F(T, V, N) = -k_B T \ln Q(T, V, N) \tag{2.10}$$

can easily be calculated. The function $Q(T, V, N)$ is called the *canonical partition function*, and k_B is the *Boltzmann constant*. This macroscopic partition function can be calculated from the *molecular partition functions* of the constituting molecules as

$$Q = \prod_{k=1}^{N} q_k, \tag{2.11}$$

where N is the total number of molecules and q_k is the (full) molecular partition function.

Assuming that the occupancy of any molecular mode is independent of the occupancy of other molecular modes, the molecular partition function can be factorised into the product of the translational, rotational, vibrational and electronic contributions:

$$q_k = q^{\text{trans}} \cdot q^{\text{rot}} \cdot q^{\text{vib}} \cdot q^{\text{el}} \tag{2.12}$$

In such cases, a suitable method to calculate the full molecular partition function is to calculate the translational, rotational, vibrational and electronic contributions separately.

The translational partition function in three dimensions is given by

$$q^{\text{trans}} = \left(\frac{2\pi m k_B T}{h^2}\right)^{3/2} \cdot V, \tag{2.13}$$

where V is the volume of the container, m the mass of the molecules and h the Planck constant. It is convenient to introduce the *thermal wavelength* Λ:

$$\Lambda = \frac{h}{\sqrt{2\pi m k T}} \tag{2.14}$$

Applying this shorthand notation, the translational partition function can be written in a compact form as:

$$q^{\text{trans}} = \frac{V}{\Lambda^3} \tag{2.15}$$

Typical values of the translational partition function are rather large; in 100 cm^3 oxygen gas at 25 °C, $\Lambda = 17.8$ pm and the value of q_{trans} is $1.773 \cdot 10^{30}$.

The rotational partition function of a molecule of general shape is the product of the partition functions related to the three independent rotational modes:

$$q^{\text{rot}} = \frac{1}{\sigma}\left(\frac{k_B T}{hc}\right)^{3/2}\sqrt{\frac{\pi}{ABC}} \qquad (2.16)$$

Here, A, B and C are the rotational constants associated to the three rotational axes, and c is the velocity of light in vacuum. The Greek letter σ denotes the *rotational symmetry factor*. Its significance is that, when the molecule rotates through $360°/\sigma$ degrees, it results in a configuration that is indistinguishable from the one that it started from, and the same configuration occurs σ times during a complete rotation. This symmetry – unlike energy degeneration – *decreases* the number of states, thus also reducing the partition function. In case of the ammonia molecule, $\sigma = 3$, while for a methane molecule, it is 12; as this molecule has four threefold rotational axes. It is also worth noting that the indistinguishable rotational state also occurs at a rotation of 180 degrees for *homonuclear* diatomic molecules; thus, the value of σ is 2 for them. A general (heteronuclear) diatomic molecule has only two (equivalent) rotational axes, thus only one rotational constant B, and its rotational partition function is simpler:

$$q^{\text{rot}}_{\text{lin}} = \frac{k_B T}{hcB} \qquad (2.17)$$

In case of a homonuclear rotor, this should be divided by two. At ambient temperatures, there are quite a lot of rotational states occupied. Thus, the value of the rotational partition function is also relatively high, at the order of magnitude of a few thousands.

The vibrational molecular partition function of a harmonic oscillator of frequency ν can be given as:

$$q^{\text{vib}} = \frac{1}{1 - e^{-\frac{h\nu}{kT}}} \qquad (2.18)$$

The vibrational partition function has the same expression for all (harmonic) normal modes. Thus, the full vibrational partition function can be written in the form

$$q^{vib} = q^{vib}(1) \cdot q^{vib}(2) \cdot \ldots \cdot q^{vib}(n), \qquad (2.19)$$

where the n different factors comprise all normal modes. A molecule constituted of K atoms can be considered as a mechanical object of K point masses; thus, it has $3K$ mechanical degrees of freedom. Three of those degrees describe translation, two (if the molecule is linear) or three (if the molecule is nonlinear) describe rotation. Thus, a linear molecule has $3K - 5$, a nonlinear multiatomic molecule has $3K - 6$ vibrational degrees of freedom, and the same number of (independent) normal vibrational modes. As vibrational excitation needs considerable energy compared to the available thermal energies at ambient temperature, the typical value of $q^{\text{vib}}(i)$ for normal modes at ambient temperature is between 1 and 3.

2.2 Transition State Theory

When calculating the electronic partition function, we should take into account the following. For most molecules, the energy of the lowest electronically excited state is high enough that at ambient and not too much higher temperatures, excited states contribute to a negligible amount to the electronic partition function. As a consequence, its value can be considered as 1. However, there are some important exceptions. There exist molecules whose electronic ground state is degenerated. The electronic partition function for them is identical to their *electronic degeneracy* g^E instead of the usual 1:

$$q^{el} = g^E \tag{2.20}$$

Molecules whose first electronic excited state is such that its energy is very close to that of the ground state are also an interesting exception. An example that exhibits both properties is the molecule NO, having two degenerate ground states and two low-lying degenerate excited states. Setting its ground-state energy to zero, its partition function is:

$$q^{el}_{NO} = 2 + 2e^{-\frac{\varepsilon^*}{kT}}, \tag{2.21}$$

where ε^* is the energy of the excited state with respect to that of the ground state.

Having calculated the contributions to the molecular partition function as described above, for a one-component gas – if the independent occupancy of states applies – the macroscopic partition function of a system of N (identical) molecules can be written in the following form:

$$Q = \frac{1}{N!} \left(q^{trans} \cdot q^{rot} \cdot q^{vib} \cdot q^{el} \right)^N \tag{2.22}$$

The division by $N!$ is necessary for the reason that gas molecules are indistinguishable from one another. As a consequence, there are as much indistinguishable states in a gas as the possible arrangement of gas molecules exchanging them with each other. The number of these permutations is exactly $N!$. If the gas has more than one component, containing N_j molecules of the j-th kind, the partition function naturally becomes

$$Q = \prod_{j=1}^{K} \frac{1}{N_j!} \left(q_j^{trans} \cdot q_j^{rot} \cdot q_j^{vib} \cdot q_j^{el} \right)^{N_j} \tag{2.23}$$

where K is the number of components. It is worth noting that the factorisability of the molecular partition functions is strictly valid only for an ideal gas. In case this approximation cannot be done within a reasonable error, the macroscopic partition function Q is usually computed using numerical simulations on canonical ensembles containing a great number of molecules. For the sake of simplicity, in the remaining part of this chapter, we shall use the ideal gas approximation.

We can calculate the chemical equilibrium constant K at constant temperature and pressure for a general reaction $\sum_{i=1}^{R} v_i A_i = 0$ using the molecular canonical partition functions the following way:

$$K = \prod_{i=1}^{R} \left(\frac{q_i^{\ominus}}{N_A}\right)^{v_i} \cdot e^{-\frac{E_0}{RT}} \qquad (2.24)$$

In this equation, $q_i^{\ominus}$ stands for the *standard molecular partition function* of component i in the *molar volume* at the standard pressure. Accordingly, the number of particles is replaced by the Avogadro constant N_A. E_0 in the exponent is the *zero-point energy of the reaction*. We should emphasise again that Eq. (2.24) is strictly valid only if the reaction mixture behaves as an ideal gas. In addition to the validity of the factorisability of the canonical partition function, when we have switched from constant temperature and constant volume to constant temperature and constant pressure, we have applied the relation between the molar free energy F_m and the molar Gibbs-potential G_m, that is, $G_m = F_m + PV_m$, and used the approximation $PV_m = RT$. This is also strictly valid only for ideal gas mixtures. However, the equation is generally valid if the fraction $q_i^{\ominus} / N_A$ is replaced by the correct molar partition function $Q_i^{\ominus}$:

$$K = \prod_{i=1}^{R} (Q_i^{\ominus})^{v_i} \cdot e^{-\frac{E_0}{RT}} \qquad (2.25)$$

E_0 in the exponent remains the zero-point energy of the reaction. In case of non-independent molecular modes, the standard molar partition function $Q_i^{\ominus}$ can be calculated by the way mentioned above: by numerical simulations.

From a chemical point of view, the previous two equations have paramount importance. If the reacting molecules are not too large, quantum chemical methods can provide the energy of all molecular modes to a good precision; thus, it is possible to calculate relevant partition functions. For large molecules, we can use spectroscopic data to calculate energies of molecular modes; thus, we can also calculate equilibrium constants. If the equilibrium constant is known from experiments for a certain reaction, the equilibrium constant for a related reaction can also be calculated using the statistical expression. In this case, we should factorise the partition function as the product of the identical contributions and the modified contributions due to changes in some reactants or products. To calculate the unknown equilibrium constant, it is sufficient to know the ratio of the modified contributions to the original ones. Equation (2.24) also has a great importance in chemical kinetics: the quasi-equilibrium transition state theory is based on this expression.

Due to the great importance of this expression, let us show the actual form of the general equation for two commonly occurring reactions. For a *bimolecular reaction* (leading to a single product) according to the stoichiometric equation

2.2 Transition State Theory

$$A + B \rightleftharpoons C, \tag{2.26}$$

the equilibrium constant reads as follows:

$$K = \frac{N_A q_C^\ominus}{q_A^\ominus q_B^\ominus} e^{-\frac{E_0}{RT}} \tag{2.27}$$

For a *unimolecular reaction* (leading to a single product) according to the stoichiometric equation

$$A \rightleftharpoons B, \tag{2.28}$$

we can write the equilibrium constant as

$$K = \frac{q_B^\ominus}{q_A^\ominus} e^{-\frac{E_0}{RT}}. \tag{2.29}$$

As we can see, the Avogadro constant does not figure in the formula if the number of the factors in the numerator and the denominator is the same.

2.2.3 Quasi-Equilibrium Transition State Theory

The underlying assumptions of this theory are the following. Let us suppose that the (overall) reaction is in equilibrium. This equilibrium is maintained by the colliding molecules AB and the atoms C – in case they have enough energy – reaching the saddle point in the valley (marked by a red asterisk in the contour map; this is the transition state), which is the highest potential state along the bottom of the valley. (Note that the progress of the reaction on the PES is not represented by the trajectory of the three nuclei – as seen on Fig. 2.3 – but by the trajectory of one single point representing the configuration of the three nuclei.) According to the assumption of quasi-equilibrium, this state is in a thermodynamic equilibrium with the reactants. If the overall reaction is in equilibrium, there is also a thermodynamic equilibrium between the transition state and the products. We can represent this overall equilibrium with the following stoichiometric scheme:

$$H_2 + H \rightleftharpoons H \cdots H \cdots H \rightleftharpoons H + H_2 \tag{2.30}$$

In the transition state, the terminal hydrogen atoms are at equal distances from the medial one, which is considerably large (94.5 pm) compared to the equilibrium distance of 76.2 pm in an H_2 molecule. Accordingly, the two bonds in this state are quite loose and they can easily break. To get to this state, the reactant molecule AB and the atom C have to expend 38.8 kJ/mol from their energy. This energy barrier

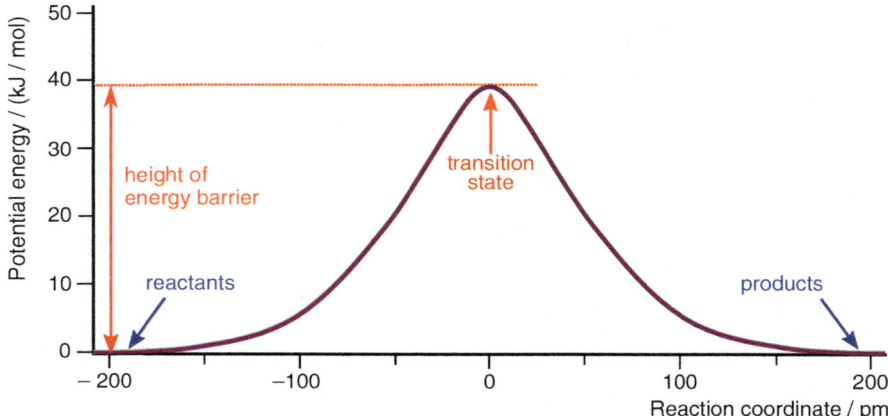

Fig. 2.6 Change of potential energy during the collinear reaction $H_2 + H \rightarrow H + H_2$. The reaction coordinate is the trajectory along the bottom of the valley in the PES. The scale is chosen so that its value is zero at the transition state; it is negative in the direction of the reactants and positive in the direction of the products

between the reactant valley and the product valley will be called simply as the *energy barrier* – conforming to the usual wording. The energy barrier can also be visualised in a two-dimensional plot if the projection to a horizontal plane of the line at the very bottom of the valley is 'straightened' into a linear coordinate, and the potential energy is plotted as a function of the configurations represented by this coordinate (Fig. 2.6).

Reaction (2.30) represents the reacting mixture when reactants and products are in thermodynamic equilibrium. The quasi-equilibrium transition state theory assumes that the first equilibrium is maintained (in a statistical thermodynamic sense) also in the case of a complete removal of the products from the reacting mixture, that is, when the backward reaction cannot occur. A second assumption is that reactants arriving at the transition state always become products. (It can be interpreted on the PES that once the reactants have 'climbed' to the saddle point, their momentum towards the product direction will move them across the potential barrier. This picture is based on a classical mechanics description, not on a quantum mechanical one.) Consequently, the unidirectional reaction model (from left to right) can be represented with the following stoichiometric scheme:

$$H_2 + H \rightleftharpoons H \cdots H \cdots H \rightarrow H + H_2 \qquad (2.31)$$

Applying these two assumptions, we can calculate the rate of reaction so that we first calculate the concentration of the transition state based on the first equilibrium, and then we calculate the rate of dissociation of the transition-state complex into the products.

2.2 Transition State Theory

Let us begin with the description of the first equilibrium. The equilibrium constant can be written the following way:

$$K^\ddagger = \frac{[\mathrm{ABC}^\ddagger]}{[\mathrm{AB}][\mathrm{C}]}, \qquad (2.32)$$

where the reactant H_2 molecule is denoted by AB, the reaction partner H atom by C, and the transition state by $\mathrm{ABC}^\ddagger$. (From here on, we shall use the usual 'double dagger' sign as a superscript to denote the transition state.) From the equilibrium constant, it is easy to express the concentration of the transition state molecules:

$$[\mathrm{ABC}^\ddagger] = K^\ddagger [\mathrm{AB}][\mathrm{C}]. \qquad (2.33)$$

Next, we shall express the equilibrium constant $K^\ddagger$ using the statistical formula:

$$K^\ddagger = \frac{N_A q^\ominus_{\mathrm{ABC}^\ddagger}}{q^\ominus_{\mathrm{AB}} q^\ominus_{\mathrm{C}}} e^{-\frac{E_0}{RT}}. \qquad (2.34)$$

Upon substitution of this into the previous equation, we get the statistical expression for the concentration of the transition state:

$$[\mathrm{ABC}^\ddagger] = \frac{N_A q^\ominus_{\mathrm{ABC}^\ddagger}}{q^\ominus_{\mathrm{AB}} q^\ominus_{\mathrm{C}}} e^{-\frac{E_0}{RT}} [\mathrm{AB}][\mathrm{C}]. \qquad (2.35)$$

Decomposition of the transition state molecules into product(s) proceeds by breaking of the bond between the atoms A and B. In the course of this bond breaking, the stretching vibrational mode of the bond A$\cdots$B becomes a translational mode, for the two atoms will not approach each other anymore – as they did during the vibrational motion in the molecule AB – but separate definitely. This dissociation can be described in two ways. We could factor out the molecular partition function of the stretching vibrational mode of A$\cdots$B from the standard molecular partition function $q^\ominus_{\mathrm{ABC}^\ddagger}$, and substitute the product of the translational partition functions of A and BC in place of it. (This can be done based on the factorisation rule that full partition functions can be given as the product of the partition functions of independent molecular modes.) We can derive the formula describing the dissociation in a simpler way taking into account that the dissociation proceeds with the frequency of the stretching mode of the loosened A$\cdots$B bond. This loose bond further loosens during dissociation (as the atom A and the molecule BC are separating). Thus, we can assume that, by factoring out the standard molecular partition function of the stretching mode of this bond, we can apply the approximation for a low value of the vibrational energy $h\nu$ with respect to the average kinetic energy $k_\mathrm{B}T$ equipartitioned at two degrees of freedom. The vibrational molecular partition function of a harmonic oscillator of frequency ν can be given as:

$$q^V = \frac{1}{1 - e^{-\frac{h\nu}{k_B T}}} \tag{2.36}$$

Taking into account the inequality $h\nu \ll k_B T$, we can use the approximation of the exponential term for exponents much smaller than 1, which gives the result of $1 - \frac{h\nu}{k_B T}$. Substituting this into the vibrational partition function leads to the following expression:

$$q^V = \frac{1}{1 - \left(1 - \frac{h\nu}{k_B T}\right)} = \frac{k_B T}{h\nu} \tag{2.37}$$

Let us write this result in place of the partition function factored out:

$$q^{\ominus}_{ABC^{\ddagger}} = \frac{k_B T}{h\nu} q^{\ominus}_{\ddagger} \tag{2.38}$$

Here and further on, $q^{\ominus}_{\ddagger}$ denotes the *truncated* standard molecular partition function that does not contain the contribution of the vibrational mode responsible for the dissociation of the transition state. Let us substitute this product in place of the full partition function $q^{\ominus}_{ABC^{\ddagger}}$:

$$[ABC^{\ddagger}] = \frac{k_B T}{h\nu} \frac{N_A q^{\ominus}_{\ddagger}}{q^{\ominus}_{AB} q^{\ominus}_{C}} e^{-\frac{E_0}{RT}} [AB][C] \tag{2.39}$$

At the left-hand side of this equation we can see the concentration of the transition state molecules, while in the denominator of the first factor on the right side is the frequency of their dissociation. This frequency can be interpreted as the number of molecules that dissociate within unit time (in case of using SI units, within a second). Multiplying the number of transition state molecules $[ABC^{\ddagger}]$ in unit volume by the frequency ν, we get the number of molecules transformed per unit time in unit volume, which is exactly *the rate of the reaction*:

$$R = \nu[ABC^{\ddagger}] = \frac{k_B T}{h} \frac{N_A q^{\ominus}_{\ddagger}}{q^{\ominus}_{AB} q^{\ominus}_{C}} e^{-\frac{E_0}{RT}} [AB][C] \tag{2.40}$$

It is interesting to see that we have got a similar expression for the reaction rate as with the collision theory; the rate of the reaction can be written as $R = k N_A N_B$. However, the rate coefficient k is considerably different from what we have derived based on the collision theory:

2.2 Transition State Theory

$$k = \frac{k_B T}{h} \frac{N_A q_{\ddagger}^{\ominus}}{q_{AB}^{\ominus} q_C^{\ominus}} e^{-\frac{E_0}{RT}} \quad (2.41)$$

We can also discover in this expression the striking presence of the account of molecular structure; information concerning details of the structure of the reactant molecule and the transition state is represented by their molecular partition functions. The content of partition functions comprises the contribution of all molecular modes. Thus, in addition to the translation that is accounted for by the collision theory, transition state theory takes into account the rotational, vibrational and electronic properties of the reacting molecules. Due to this feature, we can expect better agreement with experimental rate coefficients than in the case of collision theory.

The quasi-equilibrium description assumes that the configuration of the nuclei passes across the energy barrier exactly at the saddle point, which is equivalent to a propagation of the configuration along the very bottom of the PES valley. The saddle point has maximal potential energy along this reaction coordinate, while minimal potential energy along any other direction. Thus, if the configuration point propagates exactly along the bottom of the valley, the highest point (from the horizontal reactant valley) along this curve can be considered as the height of the barrier.

However, it is easy to show that this cannot be the case. It is well known that molecules do vibrate even at 0 K temperature – in that case, at their vibrational ground-state level. In a relatively wide range of low temperatures above 0 K, only the ground state is populated, but this vibration is inevitable. Accordingly, the interatomic distance in molecules changes while oscillating around the equilibrium bond distance and moving back and forth on the potential energy curve of Fig. 2.2. This oscillation continuously changes the bond distance within the H_2 molecule also when it approaches the other reactant, the H atom. In a similar manner, the reaction product H_2 molecule also vibrates and continuously changes bond distance. Figure 2.7 illustrates the two cases. In the left panel, we can see the hypothetical

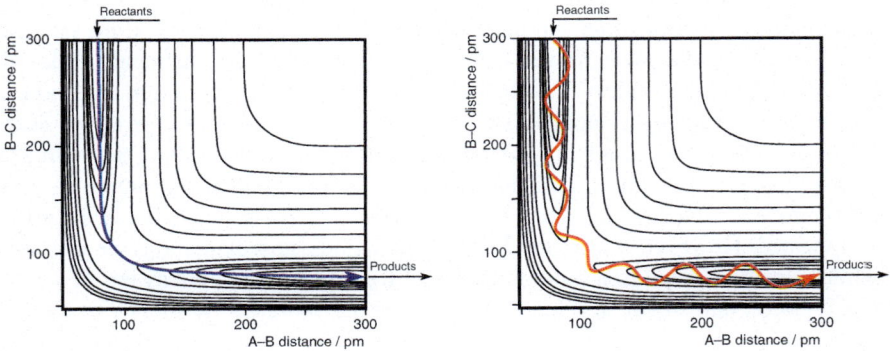

Fig. 2.7 Trajectories of the collinear reaction $H_2 + H \rightarrow H + H_2$ on the potential energy contour map. The left panel shows the hypothetical minimum potential trajectory that is located at the very bottom of the potential valley. The right panel shows the trajectory that is in accordance with the real-life H_2 molecule that oscillates at the actual vibrational frequency

non-vibrating trajectory, while the right panel shows the realistic trajectory according to oscillating H_2 molecules. We can clearly see that different trajectories with different initial vibrational phases will not cross the barrier at the saddle point but in its extended vicinity, depending on the actual phase and amplitude of the vibration. Accordingly, the 'transition state' along the realistic trajectory taking into account molecular vibration is always at a different location, but we can always find the maximum potential point along any trajectory.

To describe this more realistic procedure, we should take into account all possible trajectories with different phases and calculate the overall probability that reactants with given initial energy trespass the barrier. Performing these calculations we can see that (at constant temperature and volume), if the probability of the configuration point at the maxima of actual trajectories corresponds to the Boltzmann distribution, we get back the quasi-equilibrium result for the reaction rate. To maintain the Boltzmann distribution throughout the reaction, it is necessary that the reaction itself be somewhat slower than the energy exchange between molecules, which proceeds via collisions. In this case, the population according to the Boltzmann distribution would not be altered.

Taking into consideration the effects due to molecular vibrations, we still did not change the description concerning its validity for a collinear reaction only. In reality, reactant H_2 molecules not only oscillate but also *rotate* during the reaction, just like the newly formed H_2 molecules. Evidently, in addition to the two distances, this necessitates to take into account the angle between them, which will also continuously change during the reaction. This being an additional coordinate, the corresponding PES would be four-dimensional; not easy to visualise. According to detailed calculations, the smallest potential energy gap is experienced by the molecules having head-on (collinear) collisions. However, as every orientation is possible between gas molecules, we should also account for them. Another complication arises from the fact that classical mechanical calculations are not really valid for the description of molecular motion; we should use quantum mechanical description.

As we know, proper quantum mechanical calculations provide a true description of the behaviour of molecular motions. However – even if we apply the Born-Oppenheimer approximation and describe accordingly the reaction as if the nuclear configuration would have moved on the PES calculated for stationary, relaxed electrons around the nuclei – there are no sharp trajectories in quantum mechanics, due to the uncertainty principle. The reality is rather a temporal series of state functions, which we could only obtain by solving time-dependent Schrödinger equations. If the structure of molecules is more complex than in the simple case of the hydrogen exchange reaction, the topology of the PES also becomes more complicated. However, within the Born-Oppenheimer approximation, we can still define transition states as a multidimensional hyperplane that divides the PES into two regions: that of the reactants and that of the products. If the wave packet representing the motion of reactants crosses this hyperplane, there will be reaction and product formation. If it would not pass the hyperplane (but gets 'reflected') then there will be no reaction. In quantum mechanics – and in reality of molecular motions as well – there is also a possibility for reaction via *tunnelling*. This

2.2 Transition State Theory

tunnelling possibility plays an important role for small molecules (e.g. H_2 or D_2), especially at low temperatures.

Along the intersection of the dividing hypersurface, we can always find a saddle point that can be interpreted as the transition state. Once we have found this saddle point, we can substitute the truncated partition function $q_{\ddagger}^{\ominus}$ and the zero-point reaction energy E_0 that correspond to this state, into Eq. (2.41), and this always provides the TST reaction rate within the framework of quasi-equilibrium description. Of course, the experimental reaction rate will not necessarily agree with this calculated value, as it contains all the shortcomings of the approximations used.

Despite these shortcomings, quasi-equilibrium transition state theory is used also nowadays to describe molecular details of reactions. It reflects the qualitative properties well, and helps to understand the molecular background of reactions. As we shall discuss in the following section of this chapter, it also provides a useful basis to derive temperature and pressure dependence of reaction rates. For this reason, we shall often refer to this theory when discussing molecular details of chemical reactions.

It is worth to sum up essentials of the quasi-equilibrium TST here. This theory reflects very well molecular events happening during reactions, and can also be used for approximate quantitative estimation of reaction rates. We can generalise its formalism in terms of the following stoichiometric scheme:

$$A + B \rightleftharpoons AB^{\ddagger} \rightarrow \text{products} \quad (2.42)$$

The reaction rate coefficient related to this scheme can be calculated as follows:

$$k = \frac{k_B T}{h} \frac{N_A \, q_{\ddagger}^{\ominus}}{q_A^{\ominus} \, q_B^{\ominus}} e^{-\frac{E_0}{RT}} \quad (2.43)$$

In this expression, k_B is the Boltzmann constant, T is temperature, N_A the Avogadro constant, $q_{\ddagger}^{\ominus}$ the truncated standard molecular partition function of the transition state $AB^{\ddagger}$ not including the vibration along the reaction coordinate, $q_A^{\ominus}$ and $q_B^{\ominus}$ are the standard partition functions of the reactant molecules and E_0 is the molar energy of the activation reaction (i.e. that of the formation of $AB^{\ddagger}$ from A and B) at 0 K temperature. Accordingly, R is the molar gas constant.

Let us also recall the assumptions underlying the derivation of this result.

1. The Born-Oppenheimer approximation is applied to construct the PES; that is, every point on the PES is considered such that electrons are always relaxed to their equilibrium state. In other words, nuclear and electronic motions are separated.
2. Thermal equilibrium is supposed to apply; that is, the Boltzmann distribution applies all over the PES.

3. Even if there is no equilibrium between reactants and products, equilibrium conforming to the Boltzmann distribution is maintained between the reactants and the transition states.
4. Atomic configurations that have crossed the transition state in the direction of products will become products (i.e. they do not turn back from the top of the energy barrier).
5. The molecular mode responsible for the dissociation of the transition state can be separated from all other molecular modes and treated as a translational mode upon dissociation.

2.2.4 Dynamical Treatment of the Transition State Theory

As we have seen, the configuration point of the reacting system typically does not pass across the saddle point but in its vicinity, at different locations, depending on the vibrational phase of the reacting molecule(s). Accordingly, a correct account of the reaction rate can only be done by calculating the probability of the reactants to cross the barrier towards the products along all possible trajectories, and summing the contribution for all trajectories compatible with the state of the reactant molecules. To do this calculation, we should also interpret the notion of transition state somewhat differently than the saddle point itself.

Transition states in this new sense form a hyperplane in the PES diagram, whose dimension is less by one than that of the PES surface. This hyperplane is defined as above, by the property that, when reactant configurations cross it towards the product valley, they always become products. (On contour map Fig. 2.5, representing the collinear $H_2 + H \rightarrow H + H_2$ reaction, transition states are on a diagonal straight line connecting the lower left and upper right corners. It represents a two-dimensional plane on the tridimensional PES.) Transition states are thus always the highest energy points on the actual trajectory; from these points, molecules represented by the configuration point move downhills both 'forward' and 'backward' on the PES. Extending the quasi-equilibrium description of the TST we can say that, in order to calculate the reaction rate, we should calculate the concentration on the hyperplane of transition states and also the rate at which they cross the hyperplane to become products. However, this time we have to consider the dependence of the concentration on the actual position as well as the actual momentum of the molecules in that state.

To perform the calculations, let us introduce the following notation. The reaction coordinate (which is perpendicular to the hyperplane of the transition states) will be denoted by s, and the momentum along this coordinate by p_s. Let us denote the value of the s coordinate at the transition state (on the hyperplane) by $s^{\ddagger}$. We shall describe the motion of the configuration point not in a configuration space, but in the *phase space*, comprising also the momentum coordinates in addition to the spatial ones. Let us denote the probability density of the states in this phase space on the hyperplane by $f^{\ddagger}$. This probability density function has the property that it describes the

2.2 Transition State Theory

occupation by a state of a suitable quantum number n within an infinitesimal phase space element as

$$f^\ddagger ds dp_s = \frac{ds\, dp_s}{h} \frac{e^{-\frac{E}{k_B T}}}{q}. \qquad (2.44)$$

In this expression, the available number of quantum states (the *density of states*) in the phase space element $ds\, dp_s$ is $\frac{ds\, dp_s}{h}$, while the second fraction is a factor according to the Boltzmann distribution, q being the partition function of the corresponding state. The energy in the exponent at the location $s^\ddagger$ can be given the following way:

$$E = E_n(s) + \frac{p_s^2}{2\mu} + V_1(s). \qquad (2.45)$$

In this equation, $E_n(s)$ is the energy contained in the modes associated with intramolecular degrees of freedom, $\frac{p_s^2}{2\mu}$ is the translational energy (the kinetic energy along the s coordinate) and $V_1(s)$ the minimal potential energy at the intersection of the relevant s coordinate and the hypersurface. (The latter is exactly the energy of the PES at the given location.)

To calculate the reaction rate of the reactants in a state with quantum number n, we divide the elementary probability $f^\ddagger ds dp_s$ by ds (thus obtaining the density of states relative to unit length), then we multiply it by the velocity $\frac{ds}{dt}$ along the s coordinate. Finally, we have to integrate the expression thus obtained for all possible momenta that are directed from the reactants towards the products (i.e. $0 < p_s < \infty$), and sum the result for all possible quantum numbers n. The resulting expression for the rate coefficient is the following:

$$k = \sum_{\forall n} \int_0^\infty \frac{1}{hq} \frac{ds}{dt} e^{-\frac{\left(E_n(s) + \frac{p_s^2}{2\mu} + V_1^\ddagger\right)}{k_B T}} dp_s. \qquad (2.46)$$

(Here, $\forall$ is the quantifier meaning all possible n values.) Before actually performing the necessary operations, let us make a simplification by applying the relation between the derivative $\frac{ds}{dt}$ (the velocity along the s direction) and the momentum p_s:

$$\frac{d\left(\frac{p_s^2}{2\mu}\right)}{dp_s} = \frac{p_s}{\mu} = \frac{ds}{dt} \qquad (2.47)$$

We can easily see that the expression $\frac{ds}{dt} dp_s$ can be replaced by $dx = d\left(\frac{p_s^2}{2\mu}\right)$. Let us rewrite Eq. (2.46) for the rate coefficient accordingly, and extract constant factors before the integration:

$$k = \sum_{\forall n} \frac{1}{hq} e^{-\frac{E_n^\ddagger}{k_B T}} e^{-\frac{V_1^\ddagger}{k_B T}} \int_0^\infty e^{-\frac{x}{k_B T}} dx \qquad (2.48)$$

The primitive function of the integrand is $-k_B T\, e^{-\frac{x}{k_B T}}$, thus the result of the definite integration becomes $k_B T$. Let us substitute this into the expression of the rate coefficient and take into account that – except for the exponential function containing $E_n^\ddagger$ – all factors are constant (i.e. independent of the value of the quantum numbers, including also the potential energy $V_1^\ddagger$, as it is the energy of the ground-state reactants); thus, they can be factored out from the summation:

$$k = \frac{k_B T}{h} \frac{1}{q} e^{-\frac{V_1^\ddagger}{k_B T}} \sum_{\forall n} e^{-\frac{E_n^\ddagger}{k_B T}} \qquad (2.49)$$

Note that $\sum_{\forall n} e^{-\frac{E_n^\ddagger}{k_B T}}$ is identical to the partition function of the reactants located on the hypersurface, thus that of the transition state. Let us denote this partition function by $q_\ddagger$, as it does not contain the contribution related to the translational motion across the transition state hyperplane. Summing up the previous considerations, we can write the compact form of the rate coefficient obtained from the dynamical description of the TST as follows:

$$k = \frac{k_B T}{h} \frac{q_\ddagger}{q} e^{-\frac{V_1^\ddagger}{k_B T}} \qquad (2.50)$$

Considering that $q^\ddagger$ is the truncated partition function of the transition state, q is the partition function of the reactants and $V_1^\ddagger$ the ground-state (zero-point) energy of the formation of the transition state, we can conclude that this result is identical to that obtained with the quasi-equilibrium description of the TST. Obviously, at some point we 'smuggled' the condition of equilibrium into our dynamical derivation. Well, carefully observing the above derivation we can discover that the probability density function $f^\ddagger$ of the states in the phase space has been chosen according to the Boltzmann distribution. In other words, we re-introduced the assumption that there is a thermodynamic equilibrium between the reactants and the transition state molecules. Thus, we can conclude that extending the transition state from the saddle point to its wider surroundings represented by the hypersurface discussed above does not influence the results of the calculations, provided that the Boltzmann distribution is valid for the transition state.

However, there is a net gain in information if we consider that the dynamic description provides a tool to calculate the rate coefficient even if the probability density of the transition states does not follow the Boltzmann distribution. This situation can be interpreted in the way that, during the reaction, higher-energy states will become partially emptied with respect to their occupancy according to the

Boltzmann distribution. The reason for this is that molecular collisions and related energy distribution cannot compensate for the loss of energetic reactant molecules. However, if we know the actual probability density function that describes the density of states during the reaction, we can plug it in for $f^{\ddagger}$ and readily calculate the rate coefficient.

The practical problem is that determining the relevant probability density function is not an easy task. One of the possibilities is that we perform a numerical simulation of the reacting mixture by also taking into account the reaction itself. Obviously, once we make such a simulation, there is a possibility to obtain the rate coefficient directly from the simulation. Here we do not deal with such simulations but we would like to raise the awareness of the reader that – at the beginning of the twenty-first century – they are typically based on moving the configurations on a previously constructed PES. As we have seen when deriving the dynamical expression of the rate coefficient, the translational movement along the reaction coordinate has been treated in a classical way by considering exact values of momenta. Thus, most of the simulations are based on classical dynamics, but they can provide quite good results for the rate coefficient. There are sophisticated methods to mimic quantum mechanical calculations on the PES, which give better results.

Full quantum dynamical calculations do not even assume the Born-Oppenheimer approximation. They describe the dynamics of the reacting molecules by a time-dependent Schrödinger equation and calculate full trajectories that are either reactive, or not. A great number of such calculations is performed for all relevant reactant states, and the number of trajectories leading to reaction is divided by the total number of trajectories, thus approximating the probability of reaction, which is equivalent to the rate coefficient. However, the solution of a time-dependent Schrödinger equation still has severe limitations. At the beginning of the twenty-first century, reliable rate coefficients can be obtained for gas-phase reactions of reactants containing only a few atoms. For reactive systems containing not more than four (not too heavy) atoms, such rate coefficients can be even more precise than experimentally determined ones. It is expected that these limitations become less severe with the advent of more powerful computers and more efficient numerical methods in the future.

2.3 Dependence of the Rate Coefficient on Temperature and Pressure

Arrhenius[5] has proposed an exponential function he found to agree best with experimental results to describe the temperature dependence of the rate coefficient – already before the development of molecular theories to explain rates:

---

[5]Svante August Arrhenius (1859–1927) was a Swedish chemist. He discovered electrolytic dissociation for which he earned his PhD degree with the lowest grade in 1884 in Uppsala. He received

$$k = A\, e^{-\frac{E_a}{RT}}. \qquad (2.51)$$

The proposed relation has then been named after him. The *Arrhenius equation* did not tell anything about the so-called *pre-exponential factor* A, but collision theory has related it to the collision frequency of reactants, which is the reason to call it also as *frequency factor*. The other parameter in the equation, activation energy E_a, can be estimated to a reasonable precision from kinetic measurements even within a limited temperature range, while the pre-exponential factor's estimate seems to be independent of temperature and differs substantially from the collision frequency calculated based on the kinetic theory of gases. This fact indicates that collision theory can provide only a crude approximation of the rate coefficient.

For a long time, temperature-dependent kinetic measurements were evaluated graphically, using a transformation adjusted to the straight ruler:

$$\ln k = \ln A - \frac{E_a}{R}\frac{1}{T}. \qquad (2.52)$$

The corresponding $\ln k - \frac{1}{T}$ diagram, where transformed data lie along a straight line, is called the *Arrhenius plot*.

The definition of the activation energy is also based on the same relation:

$$\frac{d \ln k}{d\left(\frac{1}{T}\right)} = -\frac{E_a}{R}. \qquad (2.53)$$

Due to the fact that this relationship is only approximately valid, the parameter E_a obtained this way is usually called *apparent activation energy*.

Transition state theory (TST) provides a better description of the rate coefficient, thus a more precise temperature and pressure dependence as well. Considering the TST expression (2.41) for the rate coefficient, we can recognise that its last two factors comprise the equilibrium constant $K^\ddagger$ of the formation of the transition state from the reactants. Accordingly, we can write this expression also in the following form:

$$k = \frac{k_B T}{h} K^\ddagger. \qquad (2.54)$$

Recalling the relation to $\Delta G^\ddagger$ of this equilibrium constant from thermodynamics

$$K^\ddagger = e^{-\frac{\Delta G^\ddagger}{RT}}, \qquad (2.55)$$

we can substitute it into the expression of k:

---

the Nobel Prize for the same achievement in 1903. He proposed the equation for the temperature dependence of the rate coefficient in 1889 after having analysed several series of experimental data.

2.3 Dependence of the Rate Coefficient on Temperature and Pressure

$$k = \frac{k_B T}{h} e^{-\frac{\Delta G^{\ddagger}}{RT}}. \tag{2.56}$$

Considering the properties of $\Delta G^{\ddagger}$, we can make several useful conclusions. First, we can make use of the general thermodynamic relation $\Delta G = \Delta H - T\Delta S$ to get

$$k = \frac{k_B T}{h} e^{\frac{\Delta S^{\ddagger}}{R}} e^{-\frac{\Delta H^{\ddagger}}{RT}}. \tag{2.57}$$

Obviously, the first exponential factor – related to the pre-exponential factor A in the Arrhenius equation (2.51) – is accordingly a function of the *activation entropy* $\Delta S^{\ddagger}$. Second, taking the logarithm of both sides of the above equation as

$$\ln k = \ln \left(\frac{k_B T}{h}\right) + \frac{\Delta S^{\ddagger}}{R} - \frac{\Delta H^{\ddagger}}{RT}, \tag{2.58}$$

we can derive the temperature dependence of the rate coefficient. Namely, if we suppose that neither $\Delta S^{\ddagger}$ nor $\Delta H^{\ddagger}$ depends on temperature, we can express the temperature dependence of $\ln k$ as follows:[6]

$$\left(\frac{d \ln k}{dT}\right)_P = \frac{1}{T} + \frac{\Delta H^{\ddagger}}{RT^2}. \tag{2.59}$$

Changing the variable T in the derivative to $1/T$, we can rewrite the equation in a form that is similar to the van't Hoff equation related to the equilibrium constant:

$$\frac{d \ln k}{d\left(\frac{1}{T}\right)} = -\frac{\Delta H^{\ddagger} + RT}{R} \tag{2.60}$$

(Note the additive RT term not contained in the mentioned van't Hoff equation.) However, this term is barely noticeable as compared to $\Delta H^{\ddagger}$, which is typically in the range of 50 to 200 kJ/mol, while the value of RT in the vicinity of usual room temperatures is approximately 2.5 kJ/mol. Equation (2.60) also shows that, in a diagram of $\ln k$ versus $1/T$ – at least for elementary reactions – measured experimental data should align along a straight line of slope $-\frac{\Delta H^{\ddagger}+RT}{R}$. Comparing this to the linearised Arrhenius equation (2.52) we can see that the activation energy E_a is greater than the activation enthalpy $\Delta H^{\ddagger}$ by RT.

---

[6]The logarithm of the fraction $\frac{k_B T}{h}$ can be written as the sum of the logarithm of $\frac{k_B}{h}$ (which is independent of T) and the logarithm of T. The derivative of the first term is zero; thus, the result is only the derivative of the logarithm of T.

Energy and enthalpy (at constant pressure) are interrelated by $\Delta H = \Delta E + P\Delta V$; thus, we can write for the activation energy the following equation:[7]

$$E_a = \Delta H^\ddagger + RT - P\Delta V^\ddagger \qquad (2.61)$$

In case of a unimolecular reaction leading to the transition state, there is no volume change; thus, $E_a = \Delta H^\ddagger + RT$. The same is valid to a good approximation for liquid-phase reactions, as in most cases, the *activation volume* $\Delta V^\ddagger$ is negligible. In case of a bimolecular gas reaction, at least within the validity of the ideal gas law $RT = nPV$, the relation has the form $E_a = \Delta H^\ddagger$.

As a conclusion we can state that – according to the TST – the activation energy E_a estimated using the Arrhenius equation is identical to a good approximation with the activation enthalpy $\Delta H^\ddagger$. Keeping in mind that the activation enthalpy is always positive (the saddle point of the PES along the reaction coordinate is always a maximum), the increase in the rate coefficient with increasing temperature is always quite important, but it depends on the actual value of the activation enthalpy.

Let us remember that this result has been obtained supposing that neither $\Delta S^\ddagger$ nor $\Delta H^\ddagger$ depends on temperature. If we want to overcome this simplification, we can start considering the TST expression (2.41). In this formula, the fraction $\frac{k_B T}{h}$ has a first-power dependence on temperature, and it is multiplied by $\frac{N_A q_\ddagger^\ominus}{q_{AB}^\ominus q_C^\ominus}$ containing the partition functions that also have temperature as a variable at different powers in different contributions. This product is the factor of the exponential function. Thus, to calculate the exact temperature dependence of the rate coefficient, we should know the relevant partition functions exactly – which means that, in this case, we can also calculate the rate coefficient exactly and do not need to estimate it based on experimental data. However, in most of the cases we do not know the exact partition functions. In these cases, we should rely on the experimental data, taking into account the (unknown) temperature dependence of the partition functions. That is why the typical temperature-dependence formula is used in the following three-parameter form:

$$k = a\, T^n\, e^{-\frac{E}{RT}} \qquad (2.62)$$

This expression is called the *modified Arrhenius equation*.[8] Parameters a, n and E are estimated from experimental data, but a is not identical with the frequency factor A, nor is E identical with the activation energy E_a. (Except for the rare case when n proves to be significantly zero.) However, the three parameters provide enough flexibility for the function to describe temperature dependence to an excellent

---

[7]Note that the activation energy E_a is identical to the energy $\Delta E^\ddagger$ necessary to form the transition state.

[8]T in the term T^n should be understood as the temperature divided by the unit temperature; i.e. by 1 K.

2.3 Dependence of the Rate Coefficient on Temperature and Pressure

precision. Thus, this form should be used whenever the Arrhenius plot of the experimental data shows a deviation from linearity (so-called 'curved Arrhenius plots'), or equivalently, when the original exponential Arrhenius function (2.51) cannot be fitted to the experimental data without systematic errors. Typical values of the parameter a for gas reactions are of the order of 10^{11}, while for liquid-phase reactions they are even higher. The parameter n is typically between 0.5 and 1.5, but can also be much lower. The parameter E is in the usual range of the activation energies, around 50–200 kJ/mol.

To derive the pressure dependence of the rate coefficient, let us start also from Eq. (2.54). Recalling the thermodynamic relations $\ln K^{\ddagger} = -\frac{\Delta G^{\ddagger}}{RT}$ and $\left(\frac{d \Delta G^{\ddagger}}{dP}\right)_T = \Delta V^{\ddagger}$, we can write

$$\left(\frac{d \ln k}{dP}\right)_T = -\frac{\Delta V^{\ddagger}}{RT}. \tag{2.63}$$

Due to the negligibly small volume change in case of unimolecular gas reactions, we would not expect any pressure dependence of the rate coefficient. However, if the unimolecular gas reaction proceeds via thermal activation, its rate coefficient will be highly dependent on pressure, as a consequence of the Lindemann mechanism. (See Chap. 5) In case of bimolecular reactions, the transition state is formed by association; thus, the activation volume $\Delta V^{\ddagger}$ is negative, and the rate coefficient is expected to have an important exponential growth with increasing pressure. In case of liquid-phase reactions, the rate coefficient can increase or decrease according to the sign of the activation volume; however, due to small values of the activation volume, this change is typically very small.

Problems

1. In a binary gas mixture at 300 K temperature, both reactants A and B have a partial pressure of 100 Torr. The molar mass of A is 16 g/mol, and the molecule A has a diameter of 0.3 nm. The molar mass of B is 77 g/mol, and this molecule has a diameter of 0.4 nm.

    (a) Calculate the collision frequency between molecules of A and B at this temperature. (Consider the binary mixture as an ideal gas.)
    (b) What is the fraction of collisions that have enough energy to react if the activation energy of the reaction is 40 kJ/mol?
    (c) Calculate the rate coefficient of the reaction at this temperature.

    Solution: (a) Let us first convert 100 Torr to the SI pressure unit Pa: $P = 100 \text{ Torr} = 13{,}332.24 \text{ Pa}$.

    To use the formula $Z' = N_A N_B d^2 \pi \sqrt{\frac{8 k_B T}{\pi \mu}}$ for the collision frequency, we should also calculate the number density of the molecules:

$$N_A = N_B = \frac{N}{V} = \frac{P}{k_B T} = \frac{13{,}332.24 \text{ Pa}}{1.381 \times 10^{-23} \text{J/K} \cdot 300\text{K}} = 3218 \times 10^{24} \text{m}^{-3}.$$

(N is the number of molecules in volume V.) We also need to convert the molar mass to particle mass by division with the Avogadro constant, and calculate the reduced mass:

$$\mu = \frac{m_A m_B}{m_A + m_B} = \frac{16 \text{ g mol}^{-1} \cdot 77 \text{ g mol}^{-1}}{6.022 \times 10^{-23} \text{ mol}^{-1} \cdot (16+77) \text{ g mol}^{-1}} \text{ g}$$
$$= 2.1998 \times 10^{-26} \text{ kg}.$$

Now we can calculate the collision frequency:

$$Z' = 3218^2 \times 10^{48} \text{m}^{-6} \cdot 0.35^2 \times 10^{-18} \text{m}^2 \pi \sqrt{\frac{8 \cdot 1.381 \times \frac{10 \text{J}}{\text{K}} \cdot 300\text{K}}{2.1998 \times 10^{-26} \text{kg} \cdot \pi}}$$
$$= 2.76 \times 10^{33} \text{m}^{-3} \text{s}^{-1}.$$

Thus, the collision frequency is 2.76×10^{27} cm^{-3} s^{-1} in the traditional gas-kinetic unit.

(b) To calculate the fraction of molecules having more than 40 kJ/mol energy, we use Eq. (2.5): $\frac{Z'_{E_a}}{Z'} = e^{-\frac{E_a}{RT}} = e^{-\frac{40\,000 \text{ J mol}^{-1}}{8315 \text{ J K}^{-1} \text{ mol}^{-1} \cdot 300 \text{ K}}} = 1.086 \times 10^{-7}$.

(c) We can now calculate the rate coefficient using Eq. (2.8):

$$k = Z\, e^{-\frac{E_a}{RT}} = \frac{Z'}{N_A N_B} e^{-\frac{E_a}{RT}} = \frac{2.76 \times 10^{33} \text{m}^{-3} \text{s}^{-1}}{3218^2 \times 10^{48} \text{m}^{-6}} \cdot 1.086 \times 10^{-7}$$
$$= 8.268 \times 10^{-22} \text{m}^3 \text{s}^{-1}.$$

Thus, the rate coefficient is 8.268×10^{-16} cm^{3}s^{-1} in the traditional gas-kinetic unit. To convert this into the usual units, we should multiply by the Avogadro constant to have mol^{-1} unit instead of 1 over the number of particles, and multiply by 1000 to get dm^3 unit instead of m^3:

$$k = 8.268 \times 10^{-19} \text{dm}^3 \text{s}^{-1} \cdot 6.022 \times 10^{23} \text{mol}^{-1} = 4.979 \times 10^5 \text{dm}^3 \text{mol}^{-1} \text{s}^{-1}.$$

2.3 Dependence of the Rate Coefficient on Temperature and Pressure

2. Calculate the rate coefficient using TST theory of the reaction $I_2 \rightarrow 2\,I$ at 300 K temperature and 1 bar pressure. The wave number of the ground-state vibration of the I_2 molecule is $\tilde{\nu} = 223.021$ cm^{-1}. The rotational constant of the stable I_2 molecule is $B = 0.0373$ cm^{-1}, while that of the transition state is $B = 0.0364$ cm^{-1}. The zero-point energy difference between the transition state and the stable I_2 species is $E_0^\ddagger = 205$ kJ/mol.[9]

Solution: We use Eq. (2.41) adapted to the monomolecular reaction $k = \frac{k_B T}{h} \frac{q_\ddagger^\ominus}{q_{I_2}^\ominus} e^{-\frac{E_0}{RT}}$ to calculate the rate coefficient. As the expressions for the different partition function contributions are identical for both the stable reactant I_2 and the diatomic transition state $I\cdots I$, it is sufficient to write only the ratio of the parameters that are different in these expressions. We can also take into account that the vibrational mode is missing from the factor $q_\ddagger^\ominus$; thus, it should not be taken into account. To better see the terms related to each molecular mode, we can write each contribution separately and calculate their product. In this kind of calculation, the numerator of the vibrational ratio will be unit; this will not have any contribution to the product. However, the denominator then contains the full vibrational partition function. Thus the formula to use is:

$$k = \frac{k_B T}{h} \frac{q_\ddagger^{\text{trans},\ominus}}{q_{I_2}^{\text{trans},\ominus}} \frac{q_\ddagger^{\text{rot},\ominus}}{q_{I_2}^{\text{rot},\ominus}} \frac{1}{q_{I_2}^{\text{vib},\ominus}} \frac{q_\ddagger^{\text{el},\ominus}}{q_{I_2}^{\text{el},\ominus}} e^{-\frac{E_0^\ddagger}{RT}}.$$

Let us make the above-mentioned simplifications:

$$k = \frac{k_B T}{h} \frac{m_\ddagger^{3/2}}{m_{I_2}^{3/2}} \frac{1/B_\ddagger}{1/B_{I_2}} \frac{1}{\frac{1}{1-e^{-\frac{h\nu}{k_B T}}}} \cdot \frac{1}{1} \cdot e^{-\frac{E_0^\ddagger}{RT}}.$$

Let us substitute relevant values into this formula. We should convert the wave number to frequency before: 223.021 cm^{-1} = $223.021 \cdot 2.998 \times 10^{10}$ s^{-1} = 5.125×10^{12} s^{-1}.

$$k = \frac{1.381 \times 10^{-23}\,\text{J/K} \cdot 300\,\text{K}}{6.626 \times 10^{-34}\,\text{Js}} \cdot 1 \cdot \frac{0.373}{0.364} \cdot \left(1 - e^{-\frac{6.626 \times 10^{-34}\,\text{Js} \cdot 5.125 \times 10^{12}\,\text{s}^{-1}}{1.381 \times 10^{-23}\,\text{J/K} \cdot 300\,\text{K}}}\right) \cdot 1$$
$$\cdot e^{-\frac{205\,000\,\text{J mol}^{-1}}{8.315\,\text{J K}^{-1}\text{mol}^{-1} \cdot 300\,\text{K}}}\,\text{s}^{-1}.$$

---

[9] Frequency and rotational constant data are calculated using CCSD(T) def2-QZVPPD quantum chemical methods for this and the next problem. Vibrational frequencies are based on a harmonic oscillator approximation. (Calculations have been performed by Tibor Nagy.)

The result of the numerical calculations is $k = 7.327 \times 10^{-24}$ s^{-1}, a rather small value as expected from the very high zero-point activation energy.

3. Determine the isotope effect (change of the rate coefficient due to the change of an isotope) in the above reaction $I_2 \rightarrow 2\,I$ at 1500 K, using the TST theory, if the two natural iodine isotopes ^{127}I are changed to the synthetic isotopes ^{125}I (which have a half-life of their decay of 59.4 days). Vibrational wavenumbers are $\tilde{\nu}(^{127}I_2) = 223.021$ cm^{-1} and $\tilde{\nu}(^{125}I_2) = 221.256$ cm^{-1}, respectively. Rotational constants are $B(^{127}I_2) = 0.0373$ cm^{-1} and $B(^{125}I_2) = 0.0379$ cm^{-1}. Let us consider the two rotational constants of the transition states to be equal; that is, $B_{\ddagger}(^{127}I_2) = B_{\ddagger}(^{125}I_2) = 0.0364$ cm^{-1}.

Solution: The solution is related to the solution of the previous problem. Let us recall the calculation of the rate coefficient for that case:

$$k = \frac{k_B T}{h} \frac{m_{\ddagger}^{3/2}}{m_{I_2}^{3/2}} \frac{1/B_{\ddagger}}{1/B_{I_2}} \frac{1}{1 - e^{-\frac{h\nu}{k_B T}}} \cdot \frac{1}{1} \cdot e^{-\frac{E_0^{\ddagger}}{RT}}$$

$$= \frac{k_B T}{h} \frac{1/B_{\ddagger}}{1/B_{I_2}} \left(1 - e^{-\frac{h\nu}{k_B T}}\right) e^{-\frac{E_0^{\ddagger}}{RT}}.$$

The change of the energy of the species can be neglected; thus we can consider the final exponential terms in the two rate coefficients to be the same. The ratio of the two rate coefficients can then be written as

$$\frac{k_{127}}{k_{125}} = \frac{(B_{I_2}/B_{\ddagger})_{127}}{(B_{I_2}/B_{\ddagger})_{125}} \frac{1 - e^{-\frac{h\nu_{127}}{k_B T}}}{1 - e^{-\frac{h\nu_{126}}{k_B T}}}.$$

Substituting relevant parameters, we can write

$$\frac{k_{127}}{k_{125}} = \frac{0.373/0.364}{0.379/0.364} \frac{1 - e^{-\frac{6.626\times 10^{-34}\,\text{Js}\cdot 5.125\times 10^{12}\,\text{s}^{-1}}{1.381\times 10^{-23}\,\text{J K}^{-1}\text{mol}^{-1}\cdot 300\,\text{K}}}}{1 - e^{-\frac{6.626\times 10^{-34}\,\text{Js}\cdot 5.084\times 10^{12}\,\text{s}^{-1}}{1.381\times 10^{-23}\,\text{J K}^{-1}\text{mol}^{-1}\cdot 300\,\text{K}}}} = 0.992.$$

4. Let us consider the activation energy of a reaction to be 40 kJ/mol, independently of temperature. Suppose the Arrhenius equation to be valid for the temperature dependence of the rate coefficient.

   (a) How much should we increase the temperature from 300 K to double the rate of reaction?
   (b) What is the increase of the reaction rate after the same increase in temperature if the temperature dependence of the rate coefficient is best described by the modified Arrhenius equation with the following parameters: $a = 1.8 \times 10^{11}$ dm^{3}mol^{-1} s^{-1}; $n = 1.15$; $E = 56.4$ kJ/mol?

Solution: (a) Dividing the Arrhenius expression for T_2 by that for 300 K, equating the result with 2 and solving the equation we get the value of the increased temperature. The Arrhenius equation reads as $k = A\, e^{-\frac{E_a}{RT}}$. For the division, we get $\frac{k_2}{k_1} = e^{-\frac{E_a}{R}\left(\frac{1}{T_2} - \frac{1}{300\,K}\right)} = 2$. Taking the logarithm of both sides, we get $\frac{E_a}{R}\left(\frac{1}{300\,K} - \frac{1}{T_2}\right) = \ln 2$. The solution of this equation is $T_2 = 313.55$ K; this means that the reaction rate would double by a temperature increase of 13.55 K.

(b) To find the increase in the rate coefficient after the same increase in temperature, we have to do the same operations but this time using the modified Arrhenius equation $k = a\, T^n e^{-\frac{E}{RT}}$, and solving for the ratio of the rate coefficients: $\frac{k_2}{k_1} = \frac{a\, T_2^n}{a\, T_1^n} e^{-\frac{E}{R}\left(\frac{1}{T_2} - \frac{1}{T_1}\right)}$. From this we can write:

$$\ln \frac{k_2}{k_1} = n\,(\ln T_2 - \ln T_1) - \frac{E}{R}\left(\frac{1}{T_2} - \frac{1}{T_1}\right).$$

Thus, the equation to solve is the following:

$$\ln \frac{k_2}{k_1} = n\,(\ln 313.55 - \ln 300) + \frac{E}{R}\left(\frac{1}{300\,K} - \frac{1}{313.55\,K}\right).$$

The solution of this equation is $\ln \frac{k_2}{k_1} = 2.795$.

This means that, with the proper temperature dependence, the same increase of 13.55 K results in a 2.795-fold increase in the reaction rate.

Further Reading

1. Pilling MJ, Seakins PW (1995) Reaction kinetics. Oxford University Press, Oxford
2. de Paula J, Atkins PW (2014) Physical chemistry. 10th edn. Oxford University Press, Oxford
3. Silbey LJ, Alberty RA, Moungi GB (2004) Physical chemistry. 4th edn. Wiley, New York
4. Steinfeld JI, Francisco JS, Hase WL (1998) Chemical kinetics and dynamics. 2nd edn. Prentice Hall, Englewood Cliffs
5. Keszei E (2012) Chemical thermodynamics – an introduction. Springer, Berlin

Chapter 3
Formal Kinetic Description of Simple Reactions

In the preceding chapter, several theoretical models of elementary reactions are discussed that lead to the calculation of the rate coefficient. These theories suggest that the traditional form of kinetic equation describing the rate of the elementary reaction A + B → products – dating back as early as the middle of the nineteenth century – can always be written as $R = k\, N_A N_B$. (R is the reaction rate, N_A and N_B are number densities [concentrations] of the reacting species A and B. An elementary reaction is the simplest molecular event comprising only a few molecules and proceeding by the formation of only one transition state between the reactants and the products.) Similarly, if the elementary reaction is the decomposition of a single (non-thermally) activated species, then the rate equation can be written as $R = k\, N_A$. The only other – much less frequent – possibility is the reaction of three molecules (say, A, B and C) after their encounter, when the rate equation has the form $= k\, N_A N_B N_C$. We can generalise this result by writing one single equation. Replacing the number density by the more common molar concentration and writing the reaction rate also in terms of this concentration unit, the rate equation has the form

$$-\frac{1}{\nu_j}\frac{dc_j}{dt} = k \prod_{i=1}^{n} c_i, \qquad (3.1)$$

where ν_j is the stoichiometric number of the j-th component in the stoichiometric equation, c_j and c_i are actual molar concentrations and n is an integer between 1 and 3. The rate coefficient k in this equation is somewhat different from the one in the rate equations containing number densities N (usually in molecules/cm^3 unit). Let us ignore here to use another symbol for the rate coefficient when the concentration has the unit mol/dm^3, and interpret it further on as referring to the molar concentration. (The relation between the two contains the Avogadro constant, and 1000 to convert cm^3 to dm^3.) The negative sign of the time-derivative reflects the fact that reactants disappear during the course of reaction. (If we consider the stoichiometric number as being negative for the reactants, this negative sign can be dropped.) The quantity denoted by n is of course the number of reactant molecules participating in the reaction.

Rate equations having the form of Eq. (3.1) are called *mass action kinetic equations*. If we consider the stoichiometry of the reaction, we can readily realise that the concentration of the n different reactants involved do not change independently – which will help a lot to solve the rate equation. This form of the kinetic equation is typically classified according to the number of molecules participating in the molecular event of the reaction. Let us write the three stoichiometric and rate equations in the form of Eq. (3.1) for the actual cases of one, two and three molecules:

$$\text{Reaction}: \quad A \rightarrow \text{products} \tag{3.2}$$

$$\text{Rate equation}: \quad -\frac{dc}{dt} = kc \tag{3.3}$$

$$\text{Reaction}: \quad A + B \rightarrow \text{products} \tag{3.4}$$

$$\text{Rate equation}: \quad -\frac{dc_A}{dt} = -\frac{dc_B}{dt} = kc_A c_B \tag{3.5}$$

$$\text{Reaction}: \quad A + B + C \rightarrow \text{products} \tag{3.6}$$

$$\text{Rate equation}: \quad -\frac{dc_A}{dt} = -\frac{dc_B}{dt} = -\frac{dc_C}{dt} = kc_A c_B c_C \tag{3.7}$$

Observing these equations, we can see the coupling between concentration changes due to the stoichiometry, and that derivatives at the left side of the rate equations are expressed at the right side by a product, where the variables to be multiplied are not independent. Equations having this form are called *first-order homogeneous ordinary differential equations of degree n*. It is *ordinary*, as it contains the derivative of a function $c(t)$ of one single variable only. It is *first order*, as the only derivative is a first derivative. It is *homogeneous of degree n*, as there is a single function in it (apart from the derivative), which is the product of n variables; thus a polynomial of n-th degree. However, in chemistry, we call the reaction characterised by these equations as *n-th order reaction*. (This name dates back to ancient times when the *degree* of polynomials in mathematics was also called *order*. This term is no more used for the degree of polynomials.) According to this naming tradition, the three reactions listed above are *first-order*, *second-order* and *third-order* reactions, respectively.

Examples discussed above were related to molecular events happening via formation of only one transition state from the reactants and its subsequent decomposition into products. Such events are called *elementary reactions*. Their rate equation always follows the mass action kinetic law (3.1), and the number of molecules taking part in the reaction is called their *molecularity*, which – in the case of the elementary reactions – is identical to their *order*. From the point of view of molecularity, the first reaction is a *unimolecular*, the second one a *bimolecular* and the third one a *termolecular* reaction. (The latter is sometimes also called a *trimolecular* reaction, but we shall not use this term in this book.) It is worth to emphasise that the order and molecularity of the reaction are not identical categories. We shall also deal with composite reactions whose rate equation can sometimes have the form of an integer order, but molecularity is not meaningful for composite reactions that consist of several elementary reaction steps. There exist also composite reactions that have an order that is not an integer number, while molecularity always refers to integers.

It is worth mentioning why termolecular reactions are not frequent. The probability of simultaneous encounter (or, in gas phase, of simultaneous collision) of three molecules is much less frequent than that of two molecules. For this reason, a termolecular process is extremely slow, and, in many cases, there is an equivalent route with a complex mechanism that is faster, and largely masks an eventual termolecular process. Nevertheless, in liquid phase – where, due to the high density, reactant molecules are kept together for a longer time once they encountered – termolecular reactions can have an important contribution to the reaction rate.

3.1 Solution of Rate Equations of Integer-Order Reactions

As we have seen, rate equations of elementary reactions are always of integer order; thus, the title of this section could have been 'solution of rate equations of elementary reactions'. However, this would exclude the possibility for composite reactions to be of integer order, though there are plenty of examples for this behaviour.

One of the possible forms of a general integer-order reaction is the case when the initial concentrations of the reactants are identical. (This is necessarily the case if, for example, two reacting molecules are identical – though with a small difference, as it will be discussed later.) In this case, the relevant rate equation has the form

$$-\frac{dc}{dt} = kc^n, \qquad (3.8)$$

if the concentration c refers to a component with unit stoichiometric number. In mathematics, this kind of differential equation is called *separable* as the dependent variable c and the independent variable t can be separated to the opposite sides of the equation, and then both sides can readily be integrated to get an implicit function of the independent variable. The above differential equation can be written in the separated form

$$-\frac{1}{c^n} dc = k dt. \qquad (3.9)$$

Integrating both sides we can get the solution as

$$-\int \frac{1}{c^n} dc = \int k\, dt. \tag{3.10}$$

Evaluating the integrals, we can determine the primitive functions up to an undetermined additive constant. (It is enough to write one single constant; the two constants arising from the two integrations can be combined.) To find the primitive function, we can recognise that the integrands at both sides are simple power functions. On the right side – after factoring out k from the integration – we get $1 = t^0$, while on the left side, we get $\frac{1}{c^n} = c^{-n}$. Substituting their primitive functions in place of the integrals, the *general solution* is

$$-\frac{c^{1-n}}{1-n} = kt + I, \tag{3.11}$$

where the left side can be rewritten in a more transparent form:

$$\frac{1}{(n-1)\,c^{n-1}} = kt + I. \tag{3.12}$$

From this general solution, we can get the *particular solution* by finding the undetermined integration constant I with the help of the *initial conditions*. For a first-order ordinary differential equation, one initial condition is sufficient; the most convenient in this case is to give the value of the concentration at the very beginning of the reaction. Let us denote this concentration at $t = 0$ by c_o. (Further on we will call this concentration as the *initial concentration*.) Substituting $t = 0$ and $c = c_o$, we readily get the value of the integration constant as

$$I = \frac{1}{(n-1)c_o^{n-1}}. \tag{3.13}$$

Let us write this result into the general solution:

$$\frac{1}{(n-1)\,c^{n-1}} = \frac{1}{(n-1)c_o^{n-1}} + kt. \tag{3.14}$$

Multiplying both sides by $(n-1)$ we can get a somewhat simpler form

$$\frac{1}{c^{n-1}} = \frac{1}{c_o^{n-1}} + (n-1)kt, \tag{3.15}$$

from which it is straightforward to express the explicit solution $c(t)$:

3.1 Solution of Rate Equations of Integer-Order Reactions

$$c = \sqrt[n-1]{\frac{1}{\frac{1}{c_0^{n-1}} + (n-1)kt}}. \qquad (3.16)$$

We can also rewrite the ($n-1$)-th root in a power expression form, equivalent to the above:

$$c = \left(\frac{1}{\frac{1}{c_0^{n-1}} + (n-1)kt}\right)^{\frac{1}{n-1}}. \qquad (3.17)$$

It is worth noting that there is also an alternative method to solve the differential equation by evaluating *definite integrals* in accordance with the initial condition. To do this, we evaluate the right-side integral in Eq. (3.10) between the limits from the time of the initial condition $t = 0$ to an arbitrary final time t_f, and the left-side integral with respect to c from c_0 to $c(t_f)$:

$$-\int_{c_0}^{c(t_f)} \frac{1}{c^n} dc = \int_0^{t_f} k\, dt. \qquad (3.18)$$

Evaluating the integral, we get the result

$$\frac{1}{(n-1)} \left(\frac{1}{[c(t_f)]^{n-1}} - \frac{1}{c_0^{n-1}} \right) = kt_f. \qquad (3.19)$$

As this solution is valid for any final time t_f and the corresponding concentration $c(t_f)$, we can substitute t in place of t_f and c in place of $c(t_f)$, thus getting the same form of the solution as before:

$$\frac{1}{(n-1)\,c^{n-1}} - \frac{1}{(n-1)c_0^{n-1}} = kt. \qquad (3.20)$$

Some textbooks and research papers commit a severe formal error while using the definite integration by not making a difference between the dummy integration variable and the limit of the integration; in the above example they would *incorrectly* write

$$-\int_{c_0}^{c} \frac{1}{c^n} dc = \int_0^{t} k\, dt. \qquad (3.21)$$

We can easily see that this *should not* and *cannot* give the correct result we want to have, as the limit of integration always changes with the value of the integration variable. However, with the correct notation of using a different symbol for the dummy integration variable and the limit of the integration, the pencil work (especially the keyboard work) is not really simpler than in the case of evaluating the indefinite integrals and then calculating the integration constant. This procedure has the advantage that the possibility of introducing incorrect notation (and result) is avoided. Thus we shall follow the way to get the general solution first and then evaluating the integration constant to get the particular solution for the *initial value problem*.

Let us return to the solution as expressed in Eq. (3.16). It is readily seen that this solution is not applicable for all values of the reaction order n; for $n = 1$, the time-dependence disappears and the fraction 1/0 cannot be interpreted either. This frequently happens in the practice of natural sciences; the solution of a mathematical model does not always give meaningful result under all physically possible conditions. Another related interesting property is that, though the function itself can be interpreted and gives mathematically meaningful results (in case $n \neq 1$), for times $t < 0$, this does not have any physical (or chemical) meaning either, as no reaction occurs before the start of the reaction, that is, before $t = 0$; thus, the solution in this time region cannot be interpreted either. We shall return to the case of $n = 1$, but let us first explore the properties of the function given in Eq. (3.16).

The disappearance of reactants is traditionally characterised by the so-called *half-life*. By definition, this is the instance $t = t_{1/2}$, when the concentration of the reactants becomes $c = c_o/2$. It can readily be calculated by writing $c_o/2$ in place of c in the implicit solution (3.15):

$$\frac{2}{c_o^{n-1}} = \frac{1}{c_o^{n-1}} + (n-1)kt_{1/2}. \tag{3.22}$$

By rearranging we get the half-life $t_{1/2}$ as:

$$t_{1/2} = \frac{2^{n-1} - 1}{(n-1)k} \frac{1}{c_o^{n-1}}. \tag{3.23}$$

Observing this result, it is obvious that – not surprisingly – this formula of the half-life is not meaningful for first-order reactions, that is, for $n = 1$. For any $n \neq 1$, it can be seen that the sign of the first and the second coefficient is always the same, and the second coefficient depends on the initial concentration. According to this, reactions having an order greater than 1 'slow down' while proceeding, in the sense that the second time period needed to reduce the concentration by half is longer than the first time period necessary to halve the initial concentration, etc. Conversely, reactions having an order less than 1 'accelerate' while proceeding, in the sense that successive half-lives during the reaction become shorter and shorter due to the reduction in the concentration (Fig. 3.1).

3.1 Solution of Rate Equations of Integer-Order Reactions   45

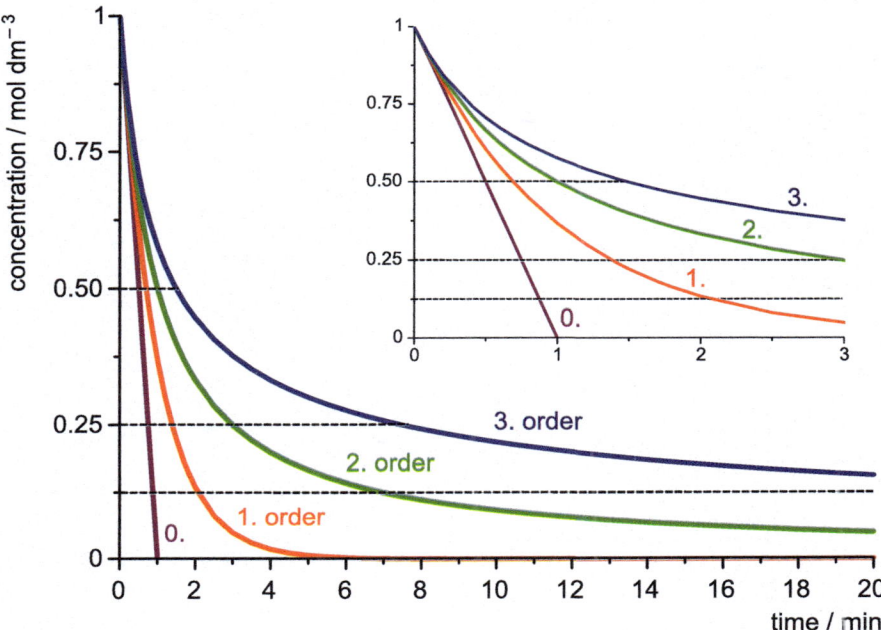

Fig. 3.1 Concentration profile of reactions of integer order in case of identical initial rates. We can see the shortening of the first, second and third half-lives in case of zero-order, as well as their lengthening in case of second- and third-order reactions. (In case of a first-order reaction, successive half-lives are identical)

It is worth discussing an aspect of the solution of the rate equations that is typically generously treated in older textbooks. Before powerful computers we use nowadays would have been widely accessible, it was much easier to determine the parameters of functions (which are c_o, n and k in case of the solution of n-th order reaction rate equations) using graphical methods. The simple and easy-to-use tool to fit functions was the (straightedge) ruler. However, as the ruler enabled only to draw straight lines, for this purpose, functions had to be 'linearised'. The linearised version of the solution of n-th order reaction rate equation is the implicit solution in the form of Eq. (3.15):

$$\frac{1}{c^{n-1}} = \frac{1}{c_0^{n-1}} + (n-1)kt. \tag{3.24}$$

We can see that, plotting the transformed $\frac{1}{c^{n-1}}$ of the concentration as a function of time t, we would get a straight line whose intercept (its value at $t = 0$) were $\frac{1}{c_0^{n-1}}$ and its slope $(n-1)k$. If we apply the ruler to draw a line across the measured points in this plot, the intersection of this line with the vertical axis (at $t = 0$) is theoretically

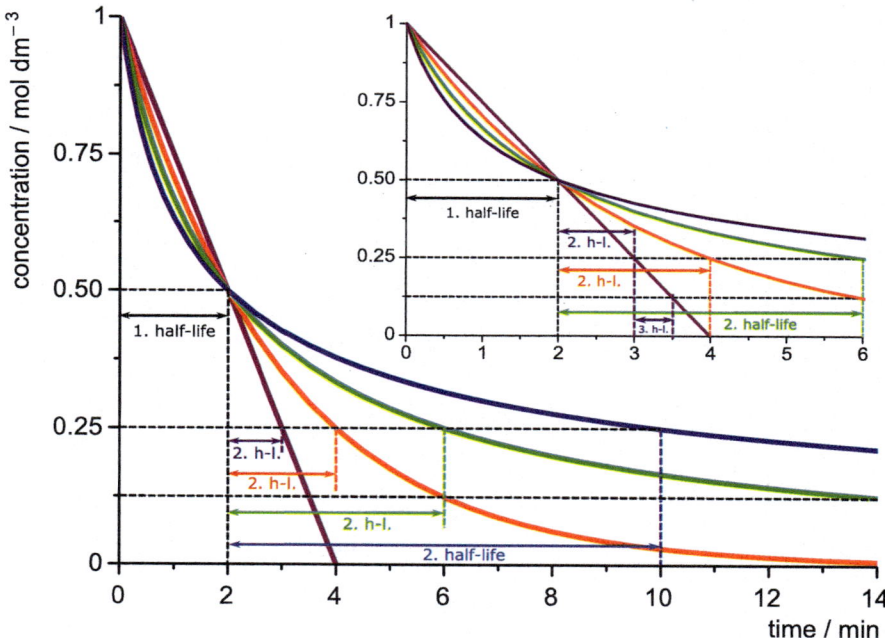

Fig. 3.2 Concentration profile of reactions of integer order in case of identical first half-lives. We can see the shortening of the second (and further) half-lives in case of zero-order, as well as their lengthening in case of second- and third-order reactions. (In case of a first-order reaction, successive half-lives are identical)

$\frac{1}{c_o^{n-1}}$ and its slope $(n-1)k$. Knowing the value of n, we could determine k from the slope.

Moreover, the order of the reaction can also be determined by the same technique of fitting a straight line with the ruler. Returning to the n-th order rate equation (3.8), let us write in place of the derivative on the left side its absolute value in the equation:

$$\left|\frac{dc}{dt}\right| = kc^n. \qquad (3.25)$$

Let us take the logarithm of both sides:

$$\log\left|\frac{dc}{dt}\right| = \log k + n \log c. \qquad (3.26)$$

We can see that plotting the transformed values $\log\left|\frac{dc}{dt}\right|$ as a function of the transformed values $\log c$, these points will be aligned along a straight line whose intercept (its value at log $c = 0$) will be log k, while its slope n. (This way, not only n but – in principle – log k could also be determined. This is the reason that this method is called the *differential method* to determine the rate coefficient.[1]) If we want to have a quick and simple picture concerning the order of reaction, the differential method may be a suitable choice. However, we should realise that numerical derivation of the discrete experimental data of the function $c(t)$ is needed, along with the logarithmic transformation of the function itself and its derivative. Furthermore, a difficulty also arises when we should allocate the derivatives (reaction rates) calculated from adjacent discrete data to some intermediate time between the two data points. These procedures thus contain some arbitrariness, and the transformations (sometimes quite heavily) distort the experimental errors. In addition, numerical derivation always increases errors. As a consequence, kinetic parameters determined this way are necessarily charged with high uncertainty.

All the problems mentioned above can easily be avoided if we do not insist using the straightedge method but perform a (nonlinear) parameter estimation based on the untransformed measured data and the explicit solution of the rate equation. (A detailed description of the proper estimation method is given in Problem 2 of this chapter.) Therefore, we shall only briefly mention the usual linearisation tricks of the concentration functions enabling (a typically inaccurate and distorted) parameter estimation with a graphical procedure using a straightedge. However, knowledge of this outdated method can help to properly understand and interpret kinetic parameters and their limitations reported in older publications, determined with graphical methods. Nowadays, with the availability of powerful computers and a great choice of suitable statistical and numerical software packages, we should prefer direct nonlinear methods to get more reliable, less distorted kinetic parameters.

Up to now, we only explored the solution of the rate equation of the reaction of order n for the case of equal initial concentration of the reactants. Let us find solutions for different initial conditions as well, and also for different values of the reaction order n. Further on in this book, we shall see that there exist some special reactions that are zero-order. (e.g. heterogeneous catalytic reactions.) The solution of their rate equation can easily be derived from the solution of the n-th order reaction discussed above. Let us begin the detailed analysis with this reaction type.

3.1.1 Zero-Order Reactions

The rate equation for a reaction that is zero-order can be written as

---

[1] In addition to this name, it is also called the *van 't Hoff method*, named after Jacobus Henricus van 't Hoff (1852–1911), the Dutch chemist who first described it. He was the first chemistry Nobel Prize winner in 1901 'for his discovery of the laws of chemical dynamics and osmotic pressure in solutions'.

$$-\frac{dc}{dt} = k, \tag{3.27}$$

for the value of the power function c^0 is always 1, independently of the actual value of c. The solution of this rate equation can easily be obtained by substituting $n = 0$ into the general solution of the n-th order reaction:

$$c = c_o - kt. \tag{3.28}$$

(Concerning the molecular interpretation of this reaction, see Sect. 6.1.)

It is obvious that this function cannot be interpreted without limitations either. Following the start of the reaction at $t = 0$, after the elapsed time $t = c_o/k$, the reactant is completely consumed and the reaction will halt; thus, negative c values after that do not have any physical meaning. To emphasise this condition, the concentration function can be given as $c = c_o - kt$, if $0 < t < c_o/k$, and zero, if $t > c_o/k$. We can also see that the unit of the rate coefficient k for a zero-order reaction is mol dm^{-3} s^{-1} in terms of molar concentration and seconds as time units – in accordance with the condition that the unit of the ratio c_o/k should be seconds. The half-life of reaction – in addition to a substitution of $n = 0$ into the general expression of the n-th order reaction – can be calculated easily by realising that the decrease in the reactant concentration is proportional to time until it reaches zero at the time c_o/k; thus it is $t_{1/2} = c_o/2k$. Obviously, consecutive half-life periods decrease by 50% each time after the concentration becomes half of the previous one. The plot of the concentration of the reactant is a straight line between c_o at $t = 0$, and zero at $t = c_o/k$.

The concentration as a function of time for the product(s) of the reaction can be obtained from the above solution relying on the stoichiometry. If the stoichiometric number of the reactant is 1 and that of a product is ν_P in the reaction reactant $\rightarrow$ products, then the concentration of the product as a function of time can be given as

$$c_P = \begin{cases} \nu_P kt, & \text{if } 0 < t < c_o/k \\ 0, & \text{if } t > c_o/k \end{cases} \tag{3.29}$$

This is easy to justify based on the relation that ν_P moles of the product are produced from 1 mol of the reactant. Thus, $c_P = \nu_P[c_o - (c_o - kt)] = \nu_P kt$.

3.1.2 First-Order Reactions

The rate equation for a first-order reaction is

$$-\frac{dc}{dt} = kc. \tag{3.30}$$

3.1 Solution of Rate Equations of Integer-Order Reactions

It is readily seen that the unit of the rate coefficient k is simply s^{-1}. The solution of this equation cannot be given by substituting $n = 1$ into the solution obtained for the general n-th order reaction, as it does not provide a meaningful result. In such cases, we should solve the actual rate equation. As this equation is also separable, after separation and insertion of the integrals, we get

$$-\int \frac{1}{c} dc = \int k \, dt \qquad (3.31)$$

Writing the primitive functions of both sides, the general solution is the following:

$$-\ln c = kt + I. \qquad (3.32)$$

Substituting the plausible initial condition ($c = c_o$ when $t = 0$), the integration constant can be obtained as $I = -\ln c_o$. Let us plug this into the above equation, and multiply it by -1 to get

$$\ln c = \ln c_o - kt. \qquad (3.33)$$

The explicit solution is readily obtained as

$$c = c_o e^{-kt}. \qquad (3.34)$$

The concentration function of the product(s) of the reaction can be obtained again relying on the stoichiometry. If ν_P moles of the product are formed from 1 mol of the reactant, based on the relation $c_P = \nu_P(c_o - c)$, we can write

$$c_P = \nu_P c_o \left(1 - e^{-kt}\right), \qquad (3.35)$$

which gives the concentration of a product with stoichiometric number ν_P as a function of time.

The validity of the above concentration functions is – in principle – not limited for times greater than zero, until infinity. However, it is worth to consider that if the reactant concentration multiplied by the volume results in a value inferior to the inverse of the Avogadro constant, there should be less than one reactant molecule in the reaction vessel. Obviously, this does not have a physical meaning; thus, in this sense, the validity of the concentration function is limited also in case of a first-order reaction. (However, the concentration would not become negative, only decrease monotonously; thus the concentration function would predict an ever smaller fraction of the last molecule if the amount of the reactant is divided by the Avogadro constant.)

The half-life of the first-order reaction has a unique property. We can calculate it by substituting $c_o/2$ in place of c:

$$\frac{c_o}{2} = c_o e^{-kt_{1/2}}. \tag{3.36}$$

Dividing both sides by c_o and taking logarithms, we get the result

$$t_{1/2} = \frac{\ln 2}{k}. \tag{3.37}$$

As can be seen, the unique property is that the half-life is independent of the initial concentration c_o. Accordingly, the concentration of the reactant for a given first-order reaction is reduced by 50% within the same time intervals, also in the case of consecutive time periods. In other words, the reactant is consumed always at the same pace; thus the reaction does not speed up nor slows down during the reaction. (It could be foreseen from the behaviour of the half-life of the n-th order reactions; below $n = 1$, the half-life decreases, above 1 it increases with decreasing initial concentration. Approaching 1 either from below or from above would lead to the same result.) (Fig. 3.3)

The conversion of the reactant at the same pace (in the sense explained above) can also be understood at the molecular level. In case of a unimolecular reaction, the probability of transformation of a molecule depends only on the internal properties of the molecule; thus it is time-independent. As a result, the number of transformed molecules is proportional to the actual number of them – as it can be seen from Eq. (3.30).

We can also notice that the implicit solution (3.33) is already appropriate for the estimation of kinetic parameters using a graphical plot and a ruler; plotting the measured log $c - t$ data in a diagram, the discrete points are found along a straight line. The slope of this line is k and its intercept is $\ln c_o$.

3.1.3 Second-Order Reactions

The rate equation for a second-order reaction *if the concentrations of two different reactants are identical* can be written in the form

$$-\frac{dc}{dt} = kc^2. \tag{3.38}$$

We can see that the SI unit of the rate coefficient k is $dm^3\,mol^{-1}\,s^{-1}$. The solution of this equation can be given by substituting $n = 2$ into the solution obtained for the general n-th order reaction:

3.1 Solution of Rate Equations of Integer-Order Reactions

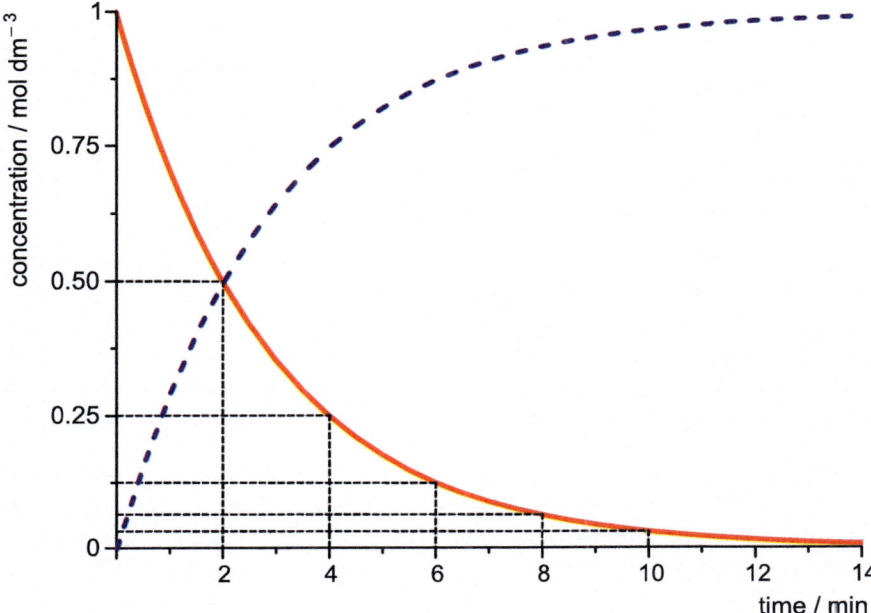

Fig. 3.3 Concentration profile of the first-order reaction A → B. It is readily seen that successive half-lives of the reactant A (solid curve) are identical. Applying this regularity, the curve can easily be traced: the actual concentration decays to its half always after the same time. The dashed curve shows the temporal evolution of the concentration of product B. Due to the stoichiometry of the reaction, the sum of the actual concentrations is identical to the sum of the initial concentrations

$$c = \frac{1}{\frac{1}{c_o} + kt}. \tag{3.39}$$

The concentration function of the product(s) of the reaction can be obtained as usual, relying on the stoichiometry. If ν_P moles of the product are produced from 1 mol of the reactant, that is, $c_P = \nu_P(c_o - c)$, we can write

$$c_P = \nu_P c_o \left(1 - \frac{1}{1 + c_o kt}\right) \tag{3.40}$$

for the concentration function of the product.

It is worth noting here also that, if the reactant concentration multiplied by the volume results in a value inferior to the inverse of the Avogadro constant, there would remain less than one reactant molecule in the reacting system; thus the validity range of the solution is limited also in this case. However, negative concentrations would never result from the concentration function.

As we have already stated, the half-life of the reactant of a second-order reaction depends on the initial concentration. Substituting $c_o/2$ in place of c, we get

$$\frac{c_o}{2} = \frac{1}{\frac{1}{c_o} + kt_{1/2}}, \qquad (3.41)$$

from which we can express the half-life:

$$t_{1/2} = \frac{1}{k\,c_o}. \qquad (3.42)$$

This reveals that the half-life of second-order reactions is inversely proportional to the initial concentration c_o. Accordingly, each consecutive reduction by 50% of the concentration takes twice as much time as the previous reduction by half. As the half-life increases during the reaction, we can say that – in this sense – the reaction 'slows down' as it proceeds. A plot of the reactant concentration as a function of time clearly reflects this tendency: the pace of the concentration change decreases during the reaction (see Figs. 3.1 and 3.2).

This decreasing pace of the reactant conversion (in the sense explained above) can also be understood at the molecular level. In case of a bimolecular reaction, the probability of transformation of reactant molecules depends not only on the internal molecular properties, but also on the probability of their encounter. The number of encounters – in case of equal concentrations c of the two reactants – is proportional to the square of the number of reactant molecules, as it can be seen in Eq. (3.38). Accordingly, the time needed to halve the number of reactant molecules is increasing, proportionally to the inverse of their initial number. (The same arguments were used to derive the results concerning the collision theory of bimolecular reactions; see Sect. 2.1.)

To get a formula for using a ruler to estimate kinetic parameters graphically, we can start from Eq. (3.39). Taking the reciprocal of both sides, we can get the following equation:

$$\frac{1}{c} = \frac{1}{c_o} + kt. \qquad (3.43)$$

We can see that the inverse of the concentration as a function of time results in a plot where the discrete $\frac{1}{c} - t$ data points are aligned along a straight line. The slope of the line is the rate coefficient k and the intercept gives $\frac{1}{c_o}$.

The attentive reader might have noticed that the obtained solution of the rate equation applies only for the case when the concentrations of the two reactants are identical, but *not the reactants themselves*. When the two reacting molecules are the same, the rate equation (3.38) slightly changes, and so does its solution. Let us begin with the relevant stoichiometric equation:

$$2A \rightarrow \text{products} \qquad (3.44)$$

The proper rate equation is the following:

3.1 Solution of Rate Equations of Integer-Order Reactions

$$-\frac{dc_A}{dt} = 2kc_A^2 \tag{3.45}$$

For the sake of simplicity, let us drop the subscript A further on:

$$-\frac{dc}{dt} = 2kc^2 \tag{3.46}$$

It is worth noting that the factor 2 appears in the rate equation as the reaction rate always *refers to the stoichiometric equation*; while the rate relative to 'one mol equation' is proportional to kc^2, the rate relative to one mol *reactant* is twice as large in this case: $2kc^2$. (We would get the same result by adding two rate equations (3.38), each of which relates only to one mole reactant.) We should remember this property of the rate equation; to get the rate of change of a component, the product of the rate coefficient and the relevant concentration(s) *should be multiplied by the stoichiometric number of the component*. (In case of a unit stoichiometric number, the factor 1 is naturally not written.)

It is readily seen that this change results only in a replacement of k by $2k$. Accordingly, the relevant solution has the form

$$c = \frac{1}{\frac{1}{c_o} + 2kt}, \tag{3.47}$$

and the half-life also changes accordingly:

$$t_{1/2} = \frac{1}{2kc_o} \tag{3.48}$$

Compared to the case when two different molecules react in a second-order reaction, in case of two identical molecules, the half-life is reduced by a factor of 2. This is not surprising as it follows from the fact that the rate of disappearance is twice as fast in the latter case than in the former. The concentration of the product(s) of the reaction as a function of time can be obtained in this case by replacing the stoichiometric number ν_P by $\nu_P/2$, and writing $2k$ in place of k in Eq. (3.40) (Fig. 3.4).

The attentive reader might also have noticed that the solution of the rate equation was applied only for the cases when the initial concentrations of the two reactants were the same, and when the two reactants were chemically identical. We still have not discussed the case of two different reactants with non-identical initial concentrations. Let us recall the relevant stoichiometric equation

$$A + B \rightarrow \text{products}, \tag{3.49}$$

along with the corresponding rate equation

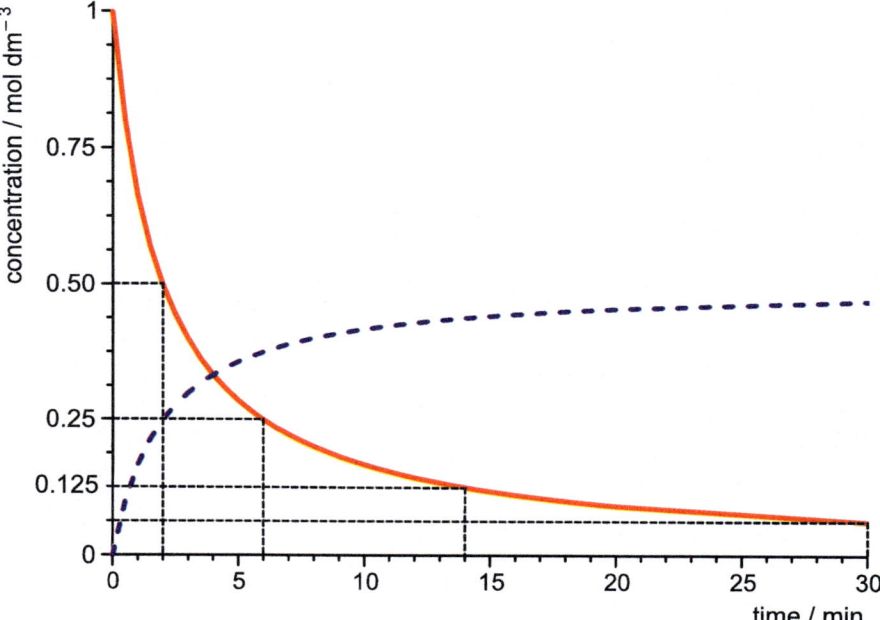

Fig. 3.4 Concentration profile of the second-order reaction 2 A → B. It is readily seen that successive half-lives of the reactant A (solid curve) are twice as long as the previous one. Applying this regularity, the curve can easily be traced: the actual concentration decays to its half always after twice the time that was necessary to halve it just before. The dashed curve shows the temporal evolution of the concentration of the product B

$$-\frac{dc_A}{dt} = -\frac{dc_B}{dt} = kc_A c_B. \tag{3.50}$$

Seemingly, there are two variables on the right side, but the stoichiometry reduces this to one underlying variable; the same amount of reactant A will always be consumed as that of the reactant B. If the volume does not change during the reaction, this relation also applies for the concentrations. Let us denote the decrease in concentration (the *progress variable*) by x, and the (different) initial concentrations of the reactants by $c_{A,o}$ and $c_{B,o}$, respectively. Using this notation, the rate equation can be written as

$$-\frac{d(c_{A,o} - x)}{dt} = -\frac{d(c_{B,o} - x)}{dt} = k(c_{A,o} - x)(c_{B,o} - x). \tag{3.51}$$

We can simplify the two time-derivatives by realising that $c_{A,o}$ and $c_{B,o}$ do not depend on time:

3.1 Solution of Rate Equations of Integer-Order Reactions

$$-\frac{d(c_{A,o} - x)}{dt} = -\frac{dc_{A,o}}{dt} + \frac{dx}{dt} = \frac{dx}{dt}. \quad (3.52)$$

We get a similar result for B as well. The differential equation to solve thus simplifies into the following form:

$$\frac{dx}{dt} = k(c_{A,o} - x)(c_{B,o} - x). \quad (3.53)$$

This differential equation is also a separable one. After separation, we get the two sides of the equation ready to integrate:

$$\int \frac{1}{(c_{A,o} - x)(c_{B,o} - x)} dx = \int k \, dt. \quad (3.54)$$

The right side provides the well-known primitive function of a zero-order power function, while we get a *rational algebraic fraction* to integrate on the left side. We can recall from calculus that the integration of this fraction can be done by resolving it into the sum of *partial fractions*:

$$\int \frac{1}{(c_{A,o} - x)(c_{B,o} - x)} dx = \int \left(\frac{1}{(c_{B,o} - c_{A,o})(c_{A,o} - x)} \right.$$
$$\left. + \frac{1}{(c_{A,o} - c_{B,o})(c_{B,o} - x)} \right) dx. \quad (3.55)$$

The integration can readily be performed resulting in the following primitive function:

$$-\ln \frac{(c_{A,o} - x)}{(c_{B,o} - c_{A,o})} - \ln \frac{(c_{B,o} - x)}{(c_{A,o} - c_{B,o})}. \quad (3.56)$$

The general solution of the differential equation including the integration constant I, after some rearrangement, can be written as

$$\frac{1}{(c_{A,o} - c_{B,o})} \ln \frac{(c_{A,o} - x)}{(c_{B,o} - x)} = kt + I. \quad (3.57)$$

Upon substitution of the initial condition ($x = 0$ at $t = 0$), the integration constant I becomes

$$I = \frac{1}{(c_{A,o} - c_{B,o})} \ln \frac{c_{A,o}}{c_{B,o}}.$$

If we plug this into the solution and make use of the identities of the logarithm function, we get

$$\frac{1}{(c_{A,o} - c_{B,o})} \ln \frac{c_{B,o}(c_{A,o} - x)}{c_{A,o}(c_{B,o} - x)} = kt. \tag{3.58}$$

Let us rewrite now the original symbols c_A and c_B in place of $c_{A,o} - x$ and $c_{B,o} - x$, and we already have the implicit solution:

$$\frac{1}{(c_{A,o} - c_{B,o})} \ln \frac{c_{B,o}\, c_A}{c_{A,o}\, c_B} = kt. \tag{3.59}$$

This result is already appropriate for the estimation of kinetic parameters using a graphical plot and a ruler; plotting the left-side transform of the concentrations as a function of time, discrete experimental points are found along a straight line across the origin, which has a slope of k.

To find the explicit solution, we should express the concentrations c_A and c_B from this equation. To do so, let us first make use of the relation $c_A = c_{A,o} - x$, and add the difference $c_{A,o} - c_{A,o}$ to $c_B = c_{B,o} - x$. This results in $c_B = c_{B,o} - x = c_{B,o} - c_{A,o} + (c_{A,o} - x)$, which is identical to $c_B = c_{B,o} - c_{A,o} + c_A$. This way we have eliminated c_B, and the remaining variable in the solution (3.59) is c_A. After some rearrangements and applying inverse logarithm (exponentiation), we can get the following result:

$$c_A = \frac{c_{B,o} - c_{A,o}}{\frac{c_{B,o}}{c_{A,o}} e^{(c_{B,o} - c_{A,o})kt} - 1}. \tag{3.60}$$

It is easy to see that – for symmetry reasons (the role of A and B can be interchanged) – the time-dependence of the concentration of B is analogous to this:

$$c_B = \frac{c_{A,o} - c_{B,o}}{\frac{c_{A,o}}{c_{B,o}} e^{(c_{A,o} - c_{B,o})kt} - 1}. \tag{3.61}$$

The above two equations provide the explicit solution for the general rate equation of second-order reactions.

However, this solution cannot be used in all circumstances. If the initial concentrations of A and B are identical, we get an expression 0/0 for the concentration functions that cannot be interpreted. Luckily, we have the previous solutions (3.39) or (3.47) for this case.

It is worth noting that the half-life of the reaction is not unique in this case, nor can it be defined under all initial conditions. The half-life of the reactant whose initial concentration is inferior with respect to the other reactant is always meaningful. This reactant can namely be completely consumed during the reaction. The reactant with higher initial concentration can only be reduced by half if its initial concentration is less than twice the concentration of the other reactant.

3.1.4 Third-Order Reactions

The rate equation of a third-order reaction for *three different reactants with identical initial concentration* can be written in the following form:

$$-\frac{dc}{dt} = kc^3. \tag{3.62}$$

We can see that the SI unit of the rate coefficient k is $dm^6\,mol^{-2}\,s^{-1}$. The solution of this rate equation can be obtained by substituting $n = 3$ into the solution of the general n-th order rate equation (3.16) with the plausible initial condition $c = c_o$ at $t = 0$:

$$c = \sqrt{\frac{1}{\frac{1}{c_o^2} + 2kt}}. \tag{3.63}$$

The concentration function of the product(s) of the reaction can be obtained as usual, taking into account the stoichiometry. If ν_P moles of the product are produced from 1 mol of the reactant, that is, $c_P = \nu_P(c_o - c)$, we can write

$$c_P = \nu_P c_o \left(1 - \sqrt{\frac{1}{1 + \frac{2kt}{c_o^2}}}\right). \tag{3.64}$$

The half-life of the reaction can be calculated by solving the equation

$$\frac{c_o}{2} = \sqrt{\frac{1}{\frac{1}{c_o^2} + 2kt_{1/2}}}. \tag{3.65}$$

After both sides are squared and their reciprocals taken, we get the result

$$t_{1/2} = \frac{3}{2kc_o^2}. \tag{3.66}$$

This shows that the half-life of third-order reactions is inversely proportional to the square of the initial concentration c_o. Accordingly, each consecutive reduction by 50% of the concentration takes four times longer than the previous reduction by half. Thus the half-life of third-order reactions increases quite largely during the reaction, which means that – in this sense – the reaction 'slows down' significantly as it proceeds. A plot of the reactant concentration as a function of time reveals this tendency.

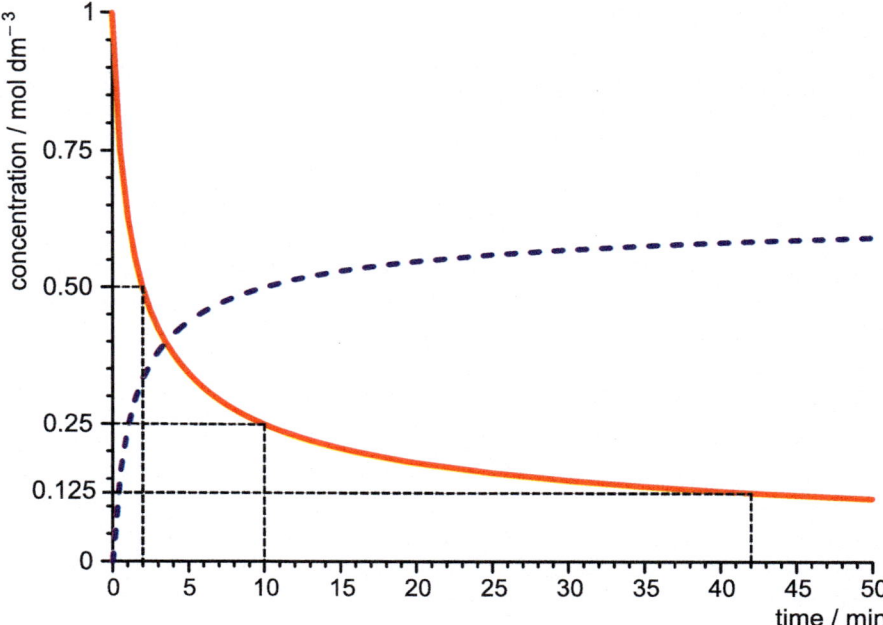

Fig. 3.5 Concentration profile of the third-order reaction A + B + C → D + E, with the initial condition that the concentrations of the reactants are equal. Successive half-lives of the reactants (solid curve) are four times as long as the previous one. Applying this regularity, the curve can easily be traced: the actual reactant concentration decays to its half always after four times longer than what was necessary to halve it just before. The dashed curve shows the temporal evolution of the concentration of the products D and E

To estimate kinetic parameters using a graphical plot and a ruler, we can derive the necessary linearised relation by starting from Eq. (3.63). After both sides are squared and their reciprocals taken, we get the following equation:

$$\frac{1}{c^2} = \frac{1}{c_o^2} + kt \tag{3.67}$$

Thus, plotting the inverse square of the concentration as a function of time, discrete experimental points are found along a straight line of slope of k and intercept $\frac{1}{c_o^2}$.

It is clear that there are reactions with different initial conditions from the above one, and we can also imagine that *three identical molecules react and the reaction rate is proportional to the third power of their concentration*. In the latter case we can follow the procedure we have discussed with second-order reactions. Starting with the relevant stoichiometric equation

3.1 Solution of Rate Equations of Integer-Order Reactions

$$3A \rightarrow \text{products}, \quad (3.68)$$

a factor of 3 will appear in the corresponding rate equation:

$$-\frac{dc_A}{dt} = 3kc_A^3. \quad (3.69)$$

The solution of this differs from the previous case in only that it contains $3k$ in place of k. Dropping the subscript A from the symbol of the concentration, the proper solution reads as follows:

$$c = \sqrt{\frac{1}{\frac{1}{c_o^2} + 6kt}}. \quad (3.70)$$

The half-life of the reaction can be calculated in a similar manner:

$$t_{1/2} = \frac{3}{6kc_o^2}. \quad (3.71)$$

If the initial concentration of the reactants is not identical, we can still discern two cases. One of them is when three different reactants take part in the reaction having different initial concentrations, and the other is when two identical reactant molecules and a third, different one, have different initial concentrations. Let us begin the description of the second case by writing the related stoichiometric equation:

$$2A + B \rightarrow \text{products}. \quad (3.72)$$

The corresponding rate equation has the following form:

$$-\frac{1}{2}\frac{dc_A}{dt} = -\frac{dc_B}{dt} = kc_A^2 c_B. \quad (3.73)$$

The two variables on the right side can also be reduced to one by realising that the amount of reactant A consumed will always be twice as much as that of reactant B. If the volume does not change during the reaction, this relation also applies for the concentrations. Let us denote the decrease in concentration (the progress variable) by x, and the (different) initial concentrations of the reactants by $c_{A,o}$ and $c_{E,o}$, respectively. Using this notation, the rate equation can be written as

$$-\frac{1}{2}\frac{d(c_{A,o} - x)}{dt} = -\frac{d(c_{B,o} - x)}{dt} = k(c_{A,o} - x)^2 (c_{B,o} - x). \quad (3.74)$$

We can simplify the two time-derivatives knowing that $c_{A,o}$ and $c_{B,o}$ do not depend on time:

$$\frac{1}{2}\frac{dx}{dt} = k(c_{A,o} - x)^2(c_{B,o} - x) \qquad (3.75)$$

The other differential equation – describing the concentration change of c_B – differs only by a factor of 2 at the left side with respect to this one. The implicit solution of this differential equation can be obtained by the method of partial fractions (similar to the second-order case) – by substituting the initial condition ($x = 0$ at $t = 0$) and re-substituting the original time-dependent concentrations c_A and c_B – yielding

$$\frac{1}{(c_{A,o} - 2c_{B,o})}\left(\frac{1}{c_{A,o}} - \frac{1}{c_A}\right) + \frac{1}{(c_{A,o} - 2c_{B,o})^2} \ln \frac{c_{B,o}\, c_A}{c_{A,o}\, c_B} = kt. \qquad (3.76)$$

This result is already appropriate for the estimation of kinetic parameters using a graphical plot and a ruler; plotting the left-side transform of the concentrations as a function of time, discrete experimental points are found along a straight line across the origin, with a slope of k. Unfortunately, the explicit solution is not known for this case. Thus, if we want to estimate kinetic parameters without relying on the graphical method, we should use numerical methods including numerical inversion of the implicit solution or numerical integration of the rate equation (3.75) to get time-dependent concentration values.

Finally, let us discuss the general third-order reaction with three different initial concentrations $c_{A,o}$, $c_{B,o}$, and $c_{C,o}$, respectively. Starting from the relevant stoichiometric equation

$$A + B + C \rightarrow \text{products}, \qquad (3.77)$$

we can write the corresponding rate equation in the following form:

$$-\frac{dc_A}{dt} = -\frac{dc_B}{dt} = -\frac{dc_C}{dt} = k c_A c_B c_C. \qquad (3.78)$$

If the volume does not change during the reaction, we can express time-dependent concentrations again with the help of a single progress variable x. Using this notation, the rate equation can be rewritten as

$$-\frac{d(c_{A,o} - x)}{dt} = -\frac{d(c_{B,o} - x)}{dt} = -\frac{d(c_{C,o} - x)}{dt}$$
$$= k(c_{A,o} - x)(c_{B,o} - x)(c_{C,o} - x). \qquad (3.79)$$

We can simplify the time-derivatives by making use of the time-independence of the initial concentrations and get the differential equation to solve:

$$\frac{dx}{dt} = k(c_{A,o} - x)(c_{B,o} - x)(c_{C,o} - x). \tag{3.80}$$

The reader might know by now that the solution of this differential equation can also be obtained by the method of partial fractions. After integration, substitution of the initial condition ($x = 0$ at $t = 0$) and re-substitution of the original time-dependent concentrations, we get the implicit solution

$$\frac{\ln \frac{c_A}{c_{A,o}}}{(c_{B,o} - c_{A,o})(c_{A,o} - c_{C,o})} + \frac{\ln \frac{c_B}{c_{B,o}}}{(c_{A,o} - c_{B,o})(c_{B,o} - c_{C,o})}$$
$$+ \frac{\ln \frac{c_C}{c_{C,o}}}{(c_{A,o} - c_{C,o})(c_{C,o} - c_{B,o})}$$
$$= kt. \tag{3.81}$$

This result is also appropriate for the estimation of kinetic parameters using a graphical plot and a ruler; plotting the left-side transform of the concentrations as a function of time, discrete experimental points are found along a straight line across the origin, with a slope of k. Unfortunately, the explicit solution is not known for this case either. Thus, if we want to estimate kinetic parameters without relying on the graphical method, we should use numerical methods, including numerical inversion of the implicit solution, or numerical integration of the rate equation to get time-dependent concentration values.

3.2 Generalisation and Extension of the Order of Reaction; Pseudo-Order

As it is mentioned in the introductory part of this chapter, there exist composite reactions that do have an order but this order is not necessarily an integer. (For elementary reactions, the order is always a positive integer.) Accordingly, we can generalise the order of reaction for any empirical rate equation, which can be written in the form

$$-\frac{dc_j}{dt} = \nu_j k \prod_{i=1}^{r} c_i^{\alpha_i}. \tag{3.82}$$

In this equation, r is the number of reactants, c_j and c_i are molar concentrations, ν_j is the stoichiometric number and the symbol α_i is called the *order of the reaction with respect to the i-th component* (or component A_i in Eq. (1.1)). Subscript j refers to the actual component whose rate is calculated, while subscript i is the dummy variable of summation running over all the components taking part in the reaction. The sum of the orders $\sum_{i=1}^{r} \alpha_i = n$ is called the *overall order* of the reaction. In relation to this name, α_i is also called the *partial order* of A_i. In a general (composite) reaction, the partial order of components as well as the overall order should not be a positive integer; it may be a rational non-integer number, or even a negative integer in case of α_i. It is worth noting that a partial order that is not a positive integer always implies a composite reaction.

As we shall see further (when discussing composite reactions), the rate equation of the majority of composite reactions cannot be written in the form of Eq. (3.82). Having this in mind we could ask why such a generalisation of the reaction order should be considered. We can have the answer also at the detailed discussion of composite reactions: when studying unknown reactions, the notion of reaction order can help in exploring kinetic properties. Kineticists often begin this exploration by 'forcing' the reaction to behave as if its rate equation would have the form of Eq. (3.82), thus at least one of its reactants would have a genuine – often integer – order. The simplest way of forcing this behaviour is to add all but one reactant in such a great excess that their concentrations during the progress of reaction remain constant to a good approximation. (This technique is sometimes referred to as *flooding*, or as *isolation*.) A frequently applied example is the case of two reactants, when the concentration of one reactant is many times that of the other.

Let us consider the case when the rate equation of the reaction of components A_1 and A_2 can be written in the second-order form

$$-\frac{dc_1}{dt} = -\frac{dc_2}{dt} = kc_1c_2. \tag{3.83}$$

For example, in case of $c_2 = 100c_1$ it is easy to see that, even after the reaction is completed, the concentration of A_2 has only been changed by 1%. With the typical error of time-dependent concentration measurements being in this order of magnitude, we do not make a big mistake by considering c_2 as a constant throughout the reaction. In that case, the product $k' = kc_2$ can also be considered constant; thus, we can rewrite the rate equation as

$$-\frac{dc_1}{dt} = k'c_1. \tag{3.84}$$

We can see that it is formally identical to a first-order rate equation. However, not being a genuine first-order reaction, it is called a *pseudo-first-order* reaction.[2]

The pseudo-first-order rate coefficient in this case has an interesting property; its usual unit is *not* s^{-1}, but $dm^3\,mol^{-1}\,s^{-1}$, according to the rate equation (3.83) or the product kc_2. Formerly, when parameter estimation had been performed using graphical methods and a ruler, it was common practice to apply the pseudo-first-order

---

[2]The Greek prefix ψευδο- has the meaning *false*, or *not a real one*.

3.2 Generalisation and Extension of the Order of Reaction; Pseudo-Order

results to determine a second-order rate coefficient. To do so, calculated linearised $\ln c_1$ vs t transformed values have been plotted at different concentrations c_2. From these diagrams, the slopes of the lines for different diagrams were determined, as discussed in Sect. 3.1.2. Pseudo-first-order rate coefficients obtained this way were plotted as a function of the corresponding concentrations c_2, and a straight line through the origin was fitted to the data points in the diagram. The slope of this line was the (graphically) estimated value of the second-order rate coefficient, according to the relation $k' = kc_2$.

In case of the graphical parameter estimations described, there is no suitable method to determine the uncertainty of the rate coefficients. Performing the described line fittings using appropriate statistical methods, we could calculate correct uncertainties of the pseudo-first-order rate coefficients, as well as the resulting second-order rate coefficient. However, it is pointless to follow this tedious method for two reasons. The first one is that the distortion of the errors due to the transformation of the original data would lead to biased results. The second one is that the statistical properties of the results obtained after the two stages (e.g. the number of the degrees of freedom of the probability distribution of the second-order rate coefficient) were unfavourable. A much simpler, more precise and statistically more sound method is to fit the concentration function (3.60) to all measured points (obtained by measuring both concentrations in a single experiment, without relying to flooding) and estimate the second-order rate coefficient as a parameter of this function, along with its uncertainty.

However, the method of flooding (or isolation) is a usual way to study unknown reactions. If it turns out, for example, that – in case of two reactants – the reaction has a pseudo-first-order kinetics for both reactants, it is a reasonable conclusion that the overall reaction is a second-order one, and we can put up a suitable experimental design to determine the second-order rate coefficient using proper nonlinear parameter estimation methods.

Problems

1. To determine the amount of an analyte B using gravimetry, it should be precipitated quantitatively. The rate coefficient of the precipitation reaction with a certain reagent A according to the equation A + B → precipitate is 0.3 dm^3 mol^{-1} s^{-1}. We require 1% precision, that is, 99% of the analyte B precipitated. What time does it take to reach this precision using stoichiometric (1:1) ratio of A and B, and various excess ratios of the reagent A? (Suppose the initial concentration of B to be 0.05 mol/dm^3.) Plot the change of concentration of the dissolved analyte B as a function of time.

Solution: To calculate the necessary reaction time for the stoichiometric ratio, let us start using Eq. (3.39) by writing $c_o/100$ into the formula:

$$\frac{c_o}{100} = \frac{1}{\frac{1}{c_o} + kt}.$$

Let us write the inverse of both sides, then rearrange to get

$$t = \frac{99}{c_o k}.$$

If reagent A is in excess, we can start using Eq. (3.58) obtained for different initial concentrations:

$$\frac{1}{(c_{A,o} - c_{B,o})} \ln \frac{c_{B,o}(c_{A,o} - x)}{c_{A,o}(c_{B,o} - x)} = kt.$$

Let us make the substitution $c_{B,o} - x = 0.01 c_{B,o}$ for the final concentration of B. The solution of this equation is $x = 0.99 c_{B,o}$. Let us write in the denominator of the argument of the logarithm $0.01 c_{B,o}$ in place of $(c_{B,o} - x)$, and in the numerator, $0.99 c_{B,o}$ in place of x:

$$\frac{1}{(c_{A,o} - c_{B,o})} \ln \frac{c_{B,o}(c_{A,o} - 0.99 c_{B,o})}{0.01 c_{A,o} c_{B,o}} = kt.$$

Rearranging and simplifying, we get the desired result:

$$t = \frac{1}{k(c_{A,o} - c_{B,o})} \ln \left(100 - 99 \frac{c_{B,o}}{c_{A,o}}\right).$$

Substituting differences and ratios for some increasing excess values, we get the following data:

A:B molar ratio	1	2	4	10	20	30	50	100
Concentration $c_{A,o}$	0.05	0.1	0.2	0.5	1	1.5	2.5	5
Time elapsed (in seconds) until $c_{B,o}$ drops to 1%	6600	261.5	96.0	33.3	16.0	10.5	6.2	3.1

Temporal evolution of the unprecipitated concentration of the analyte B is readily calculated applying solution (3.39) in case of the stoichiometric ratio, and Eq. (3.61) in case of excess of reactant A. Performing the calculations, we can plot the functions shown in the diagram above (Fig. 3.6).

In the diagram, the excess ratio is written next to the relevant curves – except the leftmost one in green colour without notation, which shows the 100-times excess of the reagent A. From the table we can see that, while the reaction in the stoichiometric ratio takes 110 min, precipitation up to 99% of the mixture with 100-times excess of the reagent A is completed within 3.1 s.

3.2 Generalisation and Extension of the Order of Reaction; Pseudo-Order

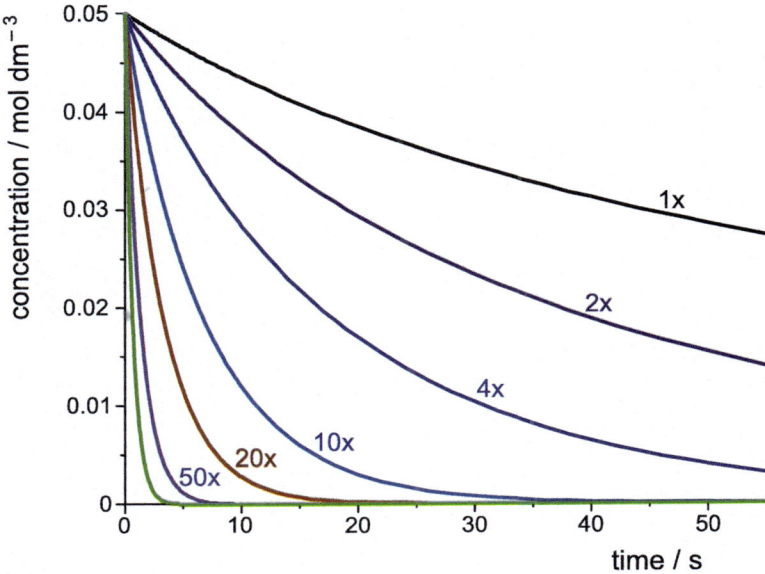

Fig. 3.6 Temporal evolution of the unprecipitated concentration of B in case of different initial excess ratio of the reactant A. Unnotated green curve shows the 100-times excess of the reagent A

It is worth noting that, in case of desired quantitative precipitation, a great excess of the reagent does not only assure a quick-enough reaction but also shifts the equilibrium towards precipitation. Due to this effect, it is worth using a great excess of reactant also in case of a rather quick reaction if the solubility constant is not small enough.

2. We have the following experimental results for the concentration of a reactant species measured during the course of a reaction to determine kinetic parameters:

time/s	1	2	3	4	5	6	7	8	9	10	11	12	13	14	15
c/mol dm^{-3}	0.564	0.371	0.361	0.285	0.288	0.250	0.219	0.203	0.199	0.198	0.209	0.163	0.168	0.170	0.141

The 'time zero' concentration could not be measured as the mixing time cannot be neglected at the timescale of seconds; thus we cannot know the actual value of concentration when the clock was started. We know that the reaction is a simple one with an integer order of 1, 2 or 3. We also know that, in case of a second- or third-order reaction, the initial concentration of all the reactants is the same c_o. Estimate the order of the reaction n, the true initial concentration c_o and the rate coefficient k.

Solution: The rate equation corresponding to the above conditions can be written as

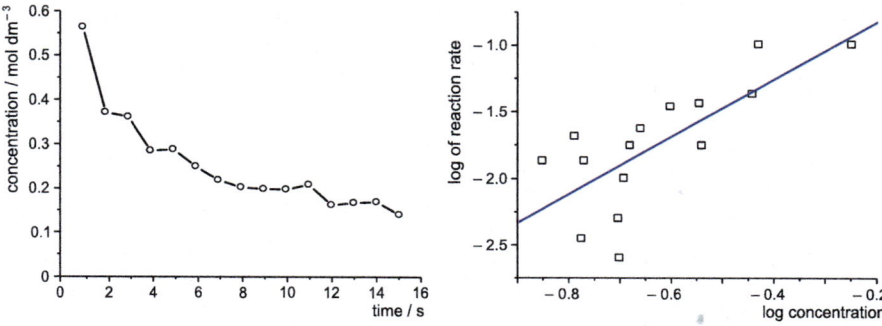

Fig. 3.7 Left panel: graphical representation of the measured data from the table. (Connecting lines only help the eye to follow the temporal sequence of data.) Right panel: Squares represent transformed data according to Eq. (3.26). The solid line is fitted to the points using a least-squares parameter estimation method

$$-\frac{dc}{dt} = kc^n. \tag{3.85}$$

The simplest way to estimate the order n of the reaction is to use the differential method. According to Eq. (3.26), transformed values of $\log \left|\frac{dc}{dt}\right|$ as a function of $\log c$ (it can be logarithm of any base) form a general straight line with the intercept (value at $\log c = 0$) $\log k$, and slope n. The right panel of Fig. 3.7 shows the transformed data (base 10 logarithm) and the line fitted to them using a least-squares parameter estimation method.

Derivatives necessary for the transformation have been calculated using the Savitzky-Golay derivation method; in this case, obtaining the derivative of a second-order polynomial calculated from the actual point and two neighbouring points, at the actual (central) point. (This method is not much different from calculating the derivative based on a smoothing with five-point moving averages and then calculating difference quotients. In case of the Savitzky-Golay method, the three points have different weights in the averaging and the derivative is 'smoother'. This derivation procedure is usually a built-in option in many statistical applications.) The 95% confidence intervals obtained for the estimated parameters of the straight line are the following: $n = 2.16 \pm 1.17$ for the slope, $\log k = -0.39 \pm 0.76$ for the intercept – the latter understood in the unit of the rate coefficient of the actual order of reaction. The 95% confidence intervals can be calculated from the estimated variance $s^2(p)$ of Student's t-distribution of the parameters p. Taking into account that we have 15 data and 2 parameters, the number of degrees of freedom of the distribution is $m = 13$, the 95% confidence interval is

$$\{p\} = p \pm t_m(0.975) \sqrt{s^2(p)}. \tag{3.86}$$

3.2 Generalisation and Extension of the Order of Reaction; Pseudo-Order

Here, $\{p\}$ is the 95% confidence interval and $t_m(0.975)$ is the value of the Student-distributed variable with m degrees of freedom, where its (cumulative) distribution function has the value of 0.975.

As we can see, the confidence interval of n contains both integers 2 and 3, but 1 is also just within the lower edge of the interval; thus we cannot decide what is the reaction order – though first order is less probable than second or third.

We can obtain the rate coefficient k from the estimated log k value; its standard deviation can be calculated taking into account the propagation of error for the univariate case:

$$s^2(f(x)) = \left(\frac{df}{dx}\right)^2 s^2(x). \tag{3.87}$$

Here, $s^2(x)$ is the variance of the random variable x, while $s^2(f(x))$ is that of the function $f(x)$. Square roots of these variances are the estimated standard deviations of x and $f(x)$, respectively.

In this actual case, the calculations are the following. The value of k is obtained as the inverse logarithm of log k: $k = 10^{\log k}$. Its standard deviation is the following:

$$s(k) = \left|\frac{d\,10^{\log k}}{d\,\log k}\right|\sqrt{s^2(\log k)} = 10^{\log k} \ln 10 \sqrt{s^2(\log k)}$$
$$= k \ln 10\, s(\log k). \tag{3.88}$$

Performing the calculations, we get 0.41 for the estimate of k and $\{-0.30; 1.12\}$ for its 95% confidence interval – in the unit of the rate coefficient of the actual order of reaction. (We cannot be surprised by the large uncertainty observing the right-panel diagram of Fig. 3.7.)

We can hope for a smaller distortion and more precise results when applying the method based on the concentration functions obtained when integrating the rate equations. These functions are different for the three different reaction orders; thus we have to estimate the parameters c_o and k in Eqs. (3.34), (3.39) and (3.63) and look for the best fit of the three model functions.

Performing the estimations and drawing the fitted functions we can see that the fit of the second- and third-order models are quite good – though there is some difference between them. The fit of the first-order model is definitely worse. Diagrams containing the fits can be seen in Figs. 3.8 and 3.9. From the diagrams we can see that, in case of the fit of the third-order model function, there is no *systematic deviation* from the observed data, which is clearly shown by the *residual errors* plotted in red colour. (Residual errors are deviations of the calculated model function values from the experimental observations. These errors can not only be visualised graphically but also be tested statistically for systematic error: *autocorrelation* or *Durbin-Watson statistics* of the residual errors quantitatively measure the extent of systematic deviation.) In case of the second-order model function, the systematic deviation is seen on the tendency of residual errors: in the first part of the course of

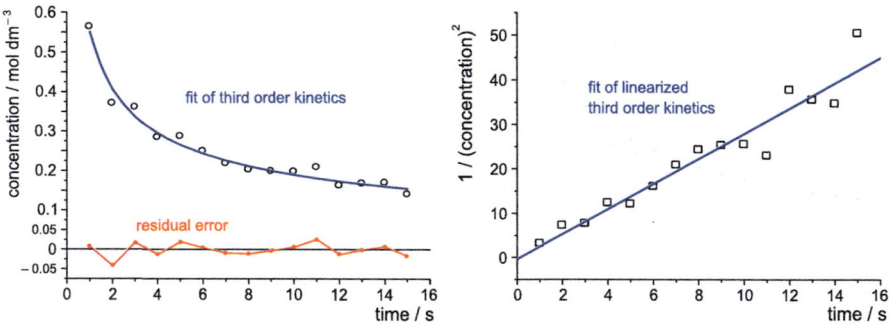

Fig. 3.8 Fit of the third-order model functions (solid blue curve/line) to the measured data (black circles/squares) shown in the table. Fitted functions are plotted according to least-squares estimation of the parameters c_o in the left panel, $\frac{1}{c_o^2}$ in the right panel and k of the functions given in Eqs. (3.63) and (3.67), respectively. Left panel: red circles fluctuating around zero show residual errors of the nonlinear fit to the original data. (Interconnecting lines only serve to see the temporal sequence.) Note that there are no systematic deviations of the residual errors: they exhibit only random fluctuations around zero, indicating a proper fit of the model function to the data. Right panel: transformed data according to the linearised model function (3.67). The blue straight line is the least-squares fit of the linearised model

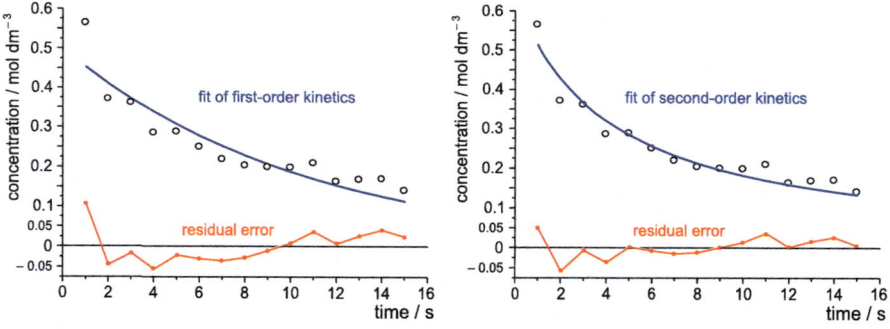

Fig. 3.9 Fits of first- and second-order model functions (solid blue curves) to the measured data (black circles) shown in the table. Fitted functions are plotted according to least-squares estimation of the parameters c_o and k of the nonlinear functions given in Eqs. (3.34) and (3.39). Red circles around zero show residual errors of the fit. (Interconnecting lines only serve to see the temporal sequence of data.) Note the systematic deviation of the residual errors: in the first part of the course of reaction they are mostly negative, while in the second part they are mostly positive. This behaviour indicates that neither first- nor second-order kinetics fits the experimental data properly

reaction they are mostly negative, while in the second part they are mostly positive. This tendency is even more pronounced in case of the first-order model function. (We can also see that the fitted first- and second-order model functions largely underestimate the measured concentration at 1 second, while in case of the third-order model, the first residual error is quite small; definitely not larger than many other residual errors. All these support that the proper order of the reaction is 3.

3.2 Generalisation and Extension of the Order of Reaction; Pseudo-Order

Table 3.1 Comparison of parameter estimation of different methods for a third-order reaction that fits the experimental results best

Method \parameter	c_o mol dm^{-3}	Confidence interval of c_o	k dm^6 mol^{-2} s	Confidence interval of k
Differential method	–	–	0.41	–0.30 to +1.12
Linearised third order	–4.36	–157 to +148	1.41	1.21–1.61
Nonlinear third order	1.34	0.60–2.08	1.35	1.19–1.51

We can see that the differential method can be used at most for obtaining preliminary information concerning the order of the reaction. The linearised model function has considerable distortions – here especially concerning the initial concentration. Nonlinear parameter estimation of the explicit concentration function gives the best result

In case of least-squares estimation, the function to minimise is the sum of squared residual differences. From these differences we can calculate the *residual error of the fit*, one of the valuable results of the parameter estimation. This residual error does not measure systematic differences between measured and calculated data but it provides information on the absolute values of average deviations of the model from the experimental data. According to the parameter estimation results, the mean value of the residuals in case of a third-order model function is 0.018 mol dm^{-3}, in case of a second-order model function 0.028 mol dm^{-3}, while in case of a first-order model function it is 0.044 mol dm^{-3}. These results also indicate that it is the third-order model function that fits best to the experimental concentration vs time data.

Results discussed above can be summarised in Table 3.1.

Observing the data in the table we can clearly see that – in case of the experimental data analysed here – the differential method is inadequate to determine the rate coefficient; moreover, the order of the reaction cannot be determined from this method either. The main reason for this failure is that input data for this method have been obtained by numerical differentiation, which always increases experimental errors. Another reason is that 'linearisation' always distorts the errors depending on the value of individual data points (see Fig. 3.7); thus smaller concentration values – which can be measured with larger relative error – have the greatest influence on the actual fit of the linearised model function. In contrast, so-called 'integrated' rate equations describing measured data without transformations provide good precision of the parameters, and the order of the reaction can also be well identified – though based on a trial-and-error procedure. As discussed in this chapter, in case of a third-order reaction, the concentration at the beginning of the course of the reaction changes rather quickly; the change significantly slowing down as the reaction proceeds. This is the reason for the great uncertainty of the initial concentration determined from the linearised method; the great distortion in case of the small concentrations' transformation leads to the undetermined initial concentration. In this particular case, the rate coefficient is quite well determined by the linearised method; however, the use of the nonlinear model function provides an unbiased rate coefficient, which also has a different value.

From these results, we can draw the general conclusion that most precise kinetic parameters can be estimated using the model function that does not distort experimental errors. For this reason, we should avoid unnecessary transformations of measured data; easy availability of computers, kinetic and statistical applications, along with their user-friendly options render any 'linearisation' or other simplifications of kinetic model functions superfluous.

We should follow this rule even in case of complicated composite reactions. Though a closed-form analytical solution of the rate equations is typically unknown for these reactions, numerical integration and parameter estimation can easily be performed based on the original measured data using computer applications discussed in Sect. 4.9.

Further Reading

1. Pilling MJ, Seakins PW (1995) Reaction kinetics. Oxford University Press, Oxford
2. de Paula J, Atkins PW (2014) Physical chemistry. 10th edn. Oxford University Press, Oxford
3. Silbey LJ, Alberty RA, Moungi GB (2004) Physical chemistry. 4th edn. Wiley, New York
4. Steinfeld JI, Francisco JS, Hase WL (1998) Chemical kinetics and dynamics. 2nd edn. Prentice Hall, Englewood Cliffs
5. Press WH, Teukolsky SA, Vetterling WT, Flannery BP (2007) Numerical recipes: the art of scientific computing. 3rd edn. Cambridge University Press, Cambridge
6. Bevington PR, Robinson DK (2002) Data reduction and error analysis for the physical sciences. 3rd edn. McGraw-Hill, New York

Chapter 4
Kinetics of Composite Reactions

Real-life chemical reactions – taking place in nature, industrial reactors, laboratory vessels or in our household – typically do not happen as one elementary reaction step; they comprise more than one collision – or more than one transition state – during the transformation of reactants to products. Products are typically formed via several elementary reactions that are connected to each other in some way. The mechanism of a *composite reaction* (or, in short, the *reaction mechanism*) is the totality of elementary steps taking place during the overall reaction along with the scheme of their connections.

This mechanism can be visualised in different ways. The most used version in reaction kinetics is to write each elementary step of the composite reaction as a stoichiometric equation. It is easily seen on the examples further in this chapter that this list of the elementary reactions already indicates the way they are connected. However, in some cases it is purposeful to depict a composite reaction in the form of a directed multigraph, where reactants and products are connected by arrows representing reactions. Another graph-like visualisation is where some reactants are connected by arrows joining to other arrows representing reactions, or products are at the head of outward branching arrows from them. There will be some examples shown of these schemes as well.

It is worth to know that the exact identification of all the elementary reactions in a composite reaction is not always done, and it is not always necessary either. In these cases, the mechanism is written (partly) in terms of so-called *mechanistic steps*, that is, in the form of stoichiometric equations that obey *mass action kinetics*. These steps can have not only first-, second- or third-order rate laws; their order can be any other number, including non-integer and negative values as well. A reaction mechanism can contain both elementary reactions and mechanistic steps. It is also possible that the reaction takes place in an open reactor, and the flow rates of entering and leaving components are written in the form of an equation similar to a stoichiometric one. These equations are said to describe (non-chemical) *sources* or *sinks* (the formation of a species from other species is called chemical source, and the disappearance via chemical reaction of a species is called chemical sink).

4.1 Coupling of Elementary Reactions

Elementary reactions taking place during the course of a composite reaction do not have a great variety of possibilities to couple. They can happen *simultaneously*: resulting in different products from the same reactants, or *consecutively*: the product(s) of a reaction becoming the reactant(s) of the next one, or *reversibly*: the reaction can proceed from the reactants towards the products but also *vice versa*.

Parallel reactions are also called *branching* or *competitive*. The name competitive refers to the fact that two or more reaction paths 'compete' for the transformation of the reactant(s). Its origin is in an illustrative representation of many physiological reactions in living organisms. The simplest example for this type of coupling is a pair of unimolecular reactions transforming a single reactant into two different products. If we want to visualise the coupling in the form of branching, we can write the following scheme:

$$A \begin{array}{c} \nearrow B \\ \searrow C \end{array} \tag{4.1}$$

If we only want to show the elementary reactions, we can write the (typographically simpler) two distinct elementary reactions.

$$\begin{array}{c} A \to B \\ A \to C \end{array} \tag{4.2}$$

As we can see, the two equations clearly show the coupling as well; the reactions are 'branched' from reactant A towards the two products B and C.

Branching reactions can be other than unimolecular. The next example illustrates a somewhat more complicated mechanism, but still branching

$$\begin{array}{c} A + B \to C + D \\ A + B \to C + E \\ A + B \to F \end{array} \tag{4.3}$$

Consecutive reactions are also called *cascade* or *serial* reactions. The simplest example for this type of coupling is also two unimolecular reactions, coupled in series.

$$A \to B \to C \tag{4.4}$$

This is a complex reaction during which an *intermediate* (or *transient*) species B is the product from the reactant A in the first step, which then reacts to give the final product C.

4.1 Coupling of Elementary Reactions

Consecutive reactions can also be much more complex than this one. In case the reactions following the first step are not unimolecular and have additional reactants, the other reactants are written above the arrow indicating reaction. A somewhat more complicated case thus could be written as follows:

$$A \xrightarrow{+B} C \to D \tag{4.5}$$

If there are more than one products in the elementary steps, it would be too much complicated to write using the above scheme; in most of the cases concerning reaction kinetic contexts, a list of the elementary reactions is given instead (in biochemical contexts, the goal of schemes is often the demonstration of large reaction networks; thus, it is more useful to depict these networks using pictorial graph representations).

Reversible reactions are also called *equilibrium* reactions. This latter name is somewhat misleading, as there is typically no equilibrium between the reactants and products of these reactions during the course of the entire composite reaction – even if they can proceed in either direction. Many reaction mechanisms contain only irreversible steps (in the forward direction only) while we know from thermodynamics that reactions can result in an equilibrium where the rates of forward and reverse reactions are equal. However, the reverse reaction can be much slower than the forward reaction, or reaction conditions can be such that reverse reactions are not important during the course of the reaction and can be neglected. Nevertheless, there are many cases where we should take into account reverse reactions as well. The simplest example for this type of coupling is also two unimolecular reactions that can proceed both in the forward and backward directions.

$$A \rightleftarrows B \tag{4.6}$$

Note that the symbol $\rightleftarrows$ is used here, rather than the usual 'equilibrium' symbol $\rightleftharpoons$, indicating that the reaction leads not necessarily to equilibrium but can proceed in either direction. In the above example, the reaction would, of course, reach equilibrium anyway if we left it for a long enough time to run. Reversible reactions can also be much more complicated; the reader can easily suggest plenty of such examples.

It is worth to emphasise that reaction mechanisms can be rather complicated, as there is a possibility for elementary reactions (or mechanistic steps) to couple in every possible way. A possible network of reactions illustrates this principle.

$$\mathrm{A+B} \begin{array}{c} \nearrow \mathrm{C} \xrightarrow{+\mathrm{F}} \mathrm{G} \begin{array}{c} \nearrow \mathrm{L} \\ \searrow \mathrm{M+N} \end{array} \\ \searrow \mathrm{D+E} \longrightarrow \mathrm{H} \rightleftharpoons \mathrm{I+J} \longrightarrow \mathrm{K} \end{array} \qquad (4.7)$$

4.2 Parallel Reactions

As a first example, let us consider the simplest case where two unimolecular reactions lead to two different products from the same reactant.

$$\mathrm{A} \begin{array}{c} \xrightarrow{k_1} \mathrm{B} \\ \xrightarrow{k_2} \mathrm{C} \end{array} \qquad (4.8)$$

The rate of a composite reaction is not written as a single differential equation but as a *system of ordinary differential equations*. To construct this, we write the rate of transformation for each component of the reacting system. For each reaction in which this component occurs, there is a term in the corresponding differential equation. A usual shorthand notation for the concentration of a species – similarly to chemical thermodynamics – is to write the chemical formula of the species within square brackets. Using this notation, we can write the following three differential equations to describe the rate of concentration changes in the above mechanism.

$$\begin{aligned} \frac{d[\mathrm{A}]}{dt} &= -k_1[\mathrm{A}] - k_2[\mathrm{A}] \\ \frac{d[\mathrm{B}]}{dt} &= k_1[\mathrm{A}] \\ \frac{d[\mathrm{C}]}{dt} &= k_2[\mathrm{A}] \end{aligned} \qquad (4.9)$$

We can see that all time-derivatives at the left-hand side of the equations are positive; thus, terms at the right-hand side are negative if the species is consumed and positive if the species is produced. Accordingly, the first equation states that component A is consumed in the reaction marked by the rate coefficient k_1 proportional to its own concentration, and it is also consumed in the reaction marked by the rate coefficient k_2, also proportional to its own concentration. The second equation tells that component B is produced in the reaction marked by the rate coefficient k_1 proportional to the concentration of A; similarly, the third equation tells that component C is produced in the reaction marked by the rate coefficient k_2, also proportional to the concentration of A.

4.2 Parallel Reactions

It is easy to see that the above system of three ordinary differential equations can readily be solved; they can be solved one by one independently, using the same method as shown in the previous chapter in case of the first-order reaction. By writing the first equation into a simpler form

$$\frac{d[A]}{dt} = -(k_1 + k_2)[A], \qquad (4.10)$$

we can readily find the solution of this first-order rate equation. Using the initial condition that at time $t = 0$, there was no product in the reaction mixture but the reactant A at a concentration denoted by $[A]_o$ (i.e. $[B]_o = 0$ and $[C]_o = 0$); the solution for [A] is

$$[A] = [A]_o e^{-(k_1+k_2)t}. \qquad (4.11)$$

This result also leads to the concentrations of the products, using the stoichiometry of the reactions. The mechanism tells us that – according to the rate equations – species B and C are always produced from A according to the ratio $k_1 : k_2$, and that, by the reaction of 1 mol of A, we always get 1 mol of the products altogether; thus, the other two solutions should read as

$$[B] = [A]_o \frac{k_1}{(k_1 + k_2)} \left(1 - e^{-(k_1+k_2)t}\right) \qquad (4.12)$$

$$[C] = [A]_o \frac{k_2}{(k_1 + k_2)} \left(1 - e^{-(k_1+k_2)t}\right). \qquad (4.13)$$

Fractions $\frac{k_1}{(k_1+k_2)}$ and $\frac{k_2}{(k_1+k_2)}$ are called *branching ratios*. The first one expresses the ratio of component B, the second one that of C in the reaction mixture relative to their sum during the entire reaction (Fig. 4.1).

Results obtained above can be extended easily for mechanisms containing any number of 'branches' if they are all first-order steps. Let us denote the common reactant by A, the products by $P_1, P_2, \ldots, P_n$. The rate equations of this mechanism with n branches are the following:

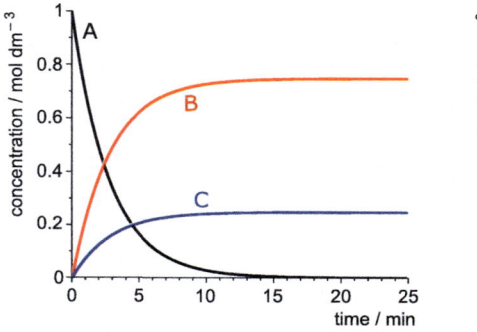
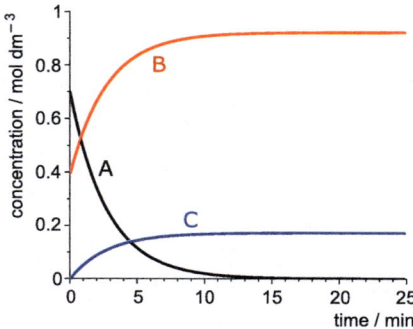

Fig. 4.1 Concentration–time diagram of the parallel (branching) reaction mechanism comprising the reactions A $\xrightarrow{k_1}$ B and A $\xrightarrow{k_2}$ C. The left panel shows the evolution of the concentrations in case when there are no products in the system at $t = 0$. The right panel shows the case when the reaction mixture contains also some product B at $t = 0$ in addition to reactant A. Note that – in general – it is not the concentration ratio [B]/[C] which is equal to the ratio k_1/k_2, but the ratio of B and C that *has been produced during the reaction*

$$\frac{d[A]}{dt} = -(k_1 + k_2 + \ldots + k_n)[A]$$
$$\frac{d[P_1]}{dt} = k_1[A]$$
$$\frac{d[P_2]}{dt} = k_2[A] \qquad (4.14)$$
$$\ldots$$
$$\frac{d[P_n]}{dt} = k_n[A]$$

Solutions of the system of differential equation (when no products are present at the beginning of the reaction) are the following:

$$[A] = [A]_o e^{-(k_1+k_2+\ldots+k_n)t}$$
$$[P_1] = [A]_o \frac{k_1}{(k_1 + k_2 + \ldots + k_n)} \left(1 - e^{-(k_1+k_2+\ldots+k_n)t}\right)$$
$$[P_2] = [A]_o \frac{k_2}{(k_1 + k_2 + \ldots + k_n)} \left(1 - e^{-(k_1+k_2+\ldots+k_n)t}\right) \qquad (4.15)$$
$$\ldots$$
$$[P_n] = [A]_o \frac{k_n}{(k_1 + k_2 + \ldots + k_n)} \left(1 - e^{-(k_1+k_2+\ldots+k_n)t}\right)$$

Accordingly, the branching ratio for the *i*th reaction can be given as $\frac{k_i}{(k_1+k_2+\ldots+k_n)}$.

Solution of more complicated branching mechanisms can be obtained in a similar manner. However, it is worth noting that solutions to complicated reaction mechanisms do not always exist in a closed form, or, as it is called in reaction kinetics, there not always exists an *analytical solution* or *explicit solution*. In these cases, we can evaluate experimental results using *numerical integration*, which will be treated further in this chapter.

4.3 Consecutive Reactions

Let us consider here also the simplest consecutive reaction shown at the beginning of this chapter; the one that consists of two consecutive unimolecular steps:

$$A \xrightarrow{k_1} B \xrightarrow{k_2} C. \tag{4.16}$$

Rate equations for this mechanism can be written according to the rules mentioned in the previous section.

$$\frac{d[A]}{dt} = -k_1[A]$$
$$\frac{d[B]}{dt} = k_1[A] - k_2[B] \tag{4.17}$$
$$\frac{d[C]}{dt} = k_2[B]$$

We can state that, again, we have one rate equation per reacting species, and the right-hand side of the differential equations contain as many terms as reactions containing the respective species. Consumption of the species is indicated by negative terms, while their formation by positive terms. The rate of change of a species is always proportional to the concentration of the *reactant* species in the respective reaction.

Solving this system of differential equations is no more as simple as it was in the case of branching reactions, but it is not that complicated either. We can see that the first equation can be solved independently of the others. Using the initial conditions that at time $t = 0$, $[A] = [A]_o$, $[B] = 0$ and $[C] = 0$, we get

$$[A] = [A]_o e^{-k_1 t}. \tag{4.18}$$

We can use this result to solve the second equation the following way. Let us substitute this solution into the rate equation of component B,

$$\frac{d[B]}{dt} = k_1[A]_o e^{-k_1 t} - k_2[B]. \tag{4.19}$$

In mathematical terms, this equation is called a *first-order non-homogeneous differential equation*. One of the well-known methods to solve such equation is to solve the *associated homogeneous differential equation* first, then *vary the constant integrating factor* in this solution.

The associated homogeneous differential equation is obtained by dropping the t-dependent term on the right-hand side

$$\frac{d[B]}{dt} = -k_2[B]. \tag{4.20}$$

The general solution of this is already familiar from the previous chapter

$$[B] = Ie^{-k_2 t}. \tag{4.21}$$

Next we should 'vary' the integrating factor I, supposing that it is a function of t. Upon substitution of the above solution into the original equation, we get:

$$\frac{d\left(Ie^{-k_2 t}\right)}{dt} = k_1[A]_o e^{-k_1 t} - k_2 Ie^{-k_2 t}. \tag{4.22}$$

When performing the derivation, the function I should cancel, only its derivative should remain, as we can see it,

$$\frac{dI}{dt}e^{-k_2 t} - Ik_2 e^{-k_2 t} = k_1[A]_o e^{-k_1 t} - k_2 Ie^{-k_2 t}. \tag{4.23}$$

Simplifying this, we get a differential equation for the function I,

$$\frac{dI}{dt} = k_1[A]_o e^{-(k_1 - k_2)t}. \tag{4.24}$$

The solution of this differential equation is the following:

$$I = \frac{k_1[A]_o}{-(k_1 - k_2)} e^{-(k_1 - k_2)t}. \tag{4.25}$$

Let us substitute this into the general solution of the homogeneous equation (4.21) to get its particular solution,

$$[B] = \frac{k_1[A]_o}{-(k_1 - k_2)} e^{-(k_1 - k_2)t} e^{-k_2 t} = \frac{k_1[A]_o}{(k_2 - k_1)} e^{-k_1 t}. \tag{4.26}$$

Finally, the sum of this and the general solution (4.21) provides the general solution of the original non-homogeneous differential equation (4.19),

4.3 Consecutive Reactions

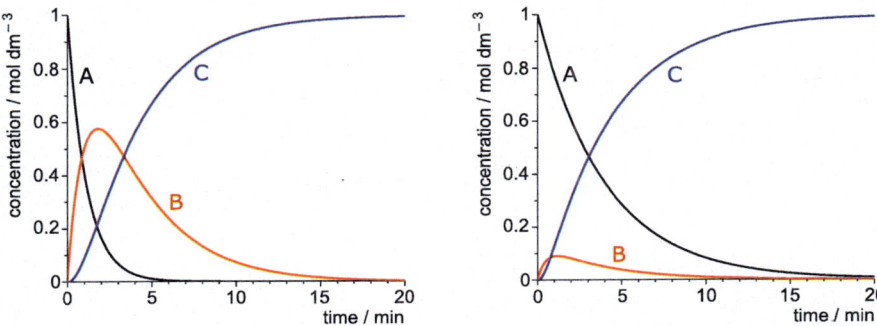

Fig. 4.2 Concentration profile of the consecutive reaction mechanism comprising the reactions A $\xrightarrow{k_1}$ B and B $\xrightarrow{k_2}$ C, with the initial condition that at $t = 0$ only component A is present in the reaction vessel. In the left panel, $k_1/k_2 = 3$, thus the intermediate B decomposes more slowly than it is produced and can accumulate temporarily during the course of reaction. In the right panel, $k_1/k_2 = 1/8$, thus the intermediate can accumulate only in a much smaller extent. During the course of reaction, the stoichiometric relation $[A] + [B] + [C] = [A]_o$ always holds

$$[B] = \frac{k_1[A]_o}{(k_2 - k_1)} e^{-k_1 t} + J e^{-k_2 t}, \tag{4.27}$$

where J is a constant factor now that can be obtained using the initial conditions. Let us use the above stated plausible initial condition that when $t = 0$, $[B] = 0$; thus, we obtain the constant

$$J = -\frac{k_1[A]_o}{(k_2 - k_1)}. \tag{4.28}$$

The last move is to substitute this into equation (4.27), and we get the solution for the concentration of B according to the above initial condition (Fig. 4.2),

$$[B] = \frac{k_1[A]_o}{(k_2 - k_1)} \left(e^{-k_1 t} - e^{-k_2 t} \right). \tag{4.29}$$

Concerning the time-dependent concentration of the species C, it is easy to get it from the stoichiometric equation (4.16), which tells us that from 1 mol of the reactant A, 1 mol of the transient B is formed; similarly, from 1 mol of the transient B, 1 mol of the product C is formed. Due to the conservation of the amount of substance – if the volume would not change during the course of reaction – the sum of the concentrations of the three species is always identical to $[A]_o$.

$$[A] + [B] + [C] = [A]_o. \tag{4.30}$$

Consequently, the time-dependent concentration of C can be written the following way:

$$[C] = [A]_o \left(1 - \frac{k_2}{(k_2 - k_1)} e^{-k_1 t} + \frac{k_1}{(k_2 - k_1)} e^{-k_2 t} \right) \tag{4.31}$$

It is worth noting that an analytical solution (i.e. a solution in a closed form) exists for all consecutive reactions comprising any number of only first-order steps (in Sect. 4.5, we shall discuss how to solve the rate equations of mechanisms comprising only first-order steps). However, if there are second-order reactions also involved in the consecutive mechanism, a solution of the rate equation in a closed form is generally not available.

4.4 Reversible Reactions

Let us consider again the simplest case where a unimolecular reaction can proceed in either direction. The stoichiometry of this mechanism is typically written in the form

$$A \underset{k_{-1}}{\overset{k_1}{\rightleftarrows}} B. \tag{4.32}$$

Let us write the rate equations for the two species,

$$\begin{aligned} \frac{d[A]}{dt} &= -k_1[A] + k_{-1}[B] \\ \frac{d[B]}{dt} &= k_1[A] - k_{-1}[B] \end{aligned} \tag{4.33}$$

First, we shall solve this system of differential equations with the special initial conditions that at $t = 0$, $[A] = [A]_o$ and $[B] = 0$. Let us use this time the initial conditions prior to the solution itself. The concentration $[A]_o$ is present either in form of A or – after having been transformed – in form of B. Consequently, the concentration of B can always be given as $[B] = [A]_o - [A]$. Let us substitute this into the equation expressing the rate of transformation of A,

$$\frac{d[A]}{dt} = -k_1[A] + k_{-1}([A]_o - [A]). \tag{4.34}$$

Observing this equation, we can see that it is also a first order non-homogeneous differential equation, similar to the one treated in the previous section; thus, we could solve it using the same method invoking the variation of the integrating factor.

4.4 Reversible Reactions

However, in this case we also have a less complicated method. Let us rearrange the equation as follows:

$$\frac{d[A]}{dt} = -(k_1 + k_{-1})[A] + k_{-1}[A]_o. \tag{4.35}$$

Let us multiply and divide at the same time the second term on the right-hand side by $(k_1 + k_{-1})$, and factor out this sum:

$$\frac{d[A]}{dt} = -(k_1 + k_{-1})\left([A] - \frac{k_{-1}}{k_1 + k_{-1}}[A]_o\right). \tag{4.36}$$

In this form we can see that it is a separable differential equation, and that – after separation – we get simple integrands at both sides of the equation,

$$\int \frac{1}{[A] - \frac{k_{-1}}{k_1+k_{-1}}[A]_o} d[A] = -\int (k_1 + k_{-1}) dt . \tag{4.37}$$

On the left-hand side, a simple fraction, while on the right-hand side, the zeroth power of the variable t has to be integrated. Writing the respective primitive functions, we can get the general solution in the following form:

$$\ln\left([A] - \frac{k_{-1}}{k_1 + k_{-1}}[A]_o\right) = -(k_1 + k_{-1})t + I. \tag{4.38}$$

The integration constant I can readily be calculated by substituting the initial conditions $t = 0$ and $[A] = [A]_o$,

$$I = \ln\left([A]_o - \frac{k_{-1}}{k_1 + k_{-1}}[A]_o\right). \tag{4.39}$$

Substituting this into the general solution and rewriting the difference of logarithms as the logarithm of the ratio of their respective arguments, we get the following particular solution,

$$\ln \frac{[A] - \frac{k_{-1}}{k_1+k_{-1}}[A]_o}{[A]_o - \frac{k_{-1}}{k_1+k_{-1}}[A]_o} = -(k_1 + k_{-1})t. \tag{4.40}$$

By taking the inverse logarithm of both sides and rearranging the equation, we get the following explicit solution,

$$[A] = [A]_o \frac{k_{-1}}{k_1 + k_{-1}} + [A]_o e^{-(k_1+k_{-1})t} - [A]_o \frac{k_{-1}}{k_1 + k_{-1}} e^{-(k_1+k_{-1})t}. \quad (4.41)$$

This form can further be simplified by factoring out $[A]_o e^{-(k_1+k_{-1})t}$ from the second and third terms on the right-hand side, then transforming the remaining $1 - \frac{k_{-1}}{k_1+k_{-1}}$ term according to $\frac{k_1+k_{-1}}{k_1+k_{-1}} - \frac{k_{-1}}{k_1+k_{-1}} = \frac{k_1}{k_1+k_{-1}}$:

$$[A] = [A]_o \frac{k_{-1}}{k_1 + k_{-1}} + [A]_o \frac{k_1}{k_1 + k_{-1}} e^{-(k_1+k_{-1})t}. \quad (4.42)$$

By further factorising, we get the following more compact form:

$$[A] = \frac{[A]_o}{k_1 + k_{-1}} \left(k_{-1} + k_1 e^{-(k_1+k_{-1})t} \right). \quad (4.43)$$

The time-dependence of the concentration of B can again be calculated using the stoichiometry, as $[B] = [A]_o - [A]$. After substituting [A] as obtained above and doing factorisation, we get:

$$[B] = [A]_o \left(1 - \frac{k_{-1}}{k_1 + k_{-1}} - \frac{k_1}{k_1 + k_{-1}} e^{-(k_1+k_{-1})t} \right). \quad (4.44)$$

Using the relation as before, that we can replace the term $1 - \frac{k_{-1}}{k_1+k_{-1}}$ according to the transformation $\frac{k_1+k_{-1}}{k_1+k_{-1}} - \frac{k_{-1}}{k_1+k_{-1}} = \frac{k_1}{k_1+k_{-1}}$, we get the following compact form:

$$[B] = [A]_o \frac{k_1}{k_1 + k_{-1}} \left(1 - e^{-(k_1+k_{-1})t} \right). \quad (4.45)$$

By doing a slightly different rearrangement, we can get another form of the same function,

$$[B] = [A]_o \frac{k_{-1}}{k_1 + k_{-1}} - [A]_o \frac{k_1}{k_1 + k_{-1}} e^{-(k_1+k_{-1})t}. \quad (4.46)$$

Comparing this form with the function (4.42) we can see that during the course of reaction, an increase in the concentration of A is accompanied by the same amount of decrease in the concentration of B and *vice versa* (this is trivially expected, but it is worth seeing deduced from the solution of the mechanism; Fig. 4.3).

As the composite reaction (4.32) is proceeding in both directions, it rarely happens in practice that at the beginning of the kinetic measurements only one of the components (A in the above calculations) is present in the reaction mixture. A more realistic initial condition is that, at $t = 0$, the concentration of A is $[A]_o$ and that of B is $[B]_o$, neither of them being zero. The solution of this problem is not more complicated than that of the previous simple case. Based on the stoichiometry, we

4.4 Reversible Reactions

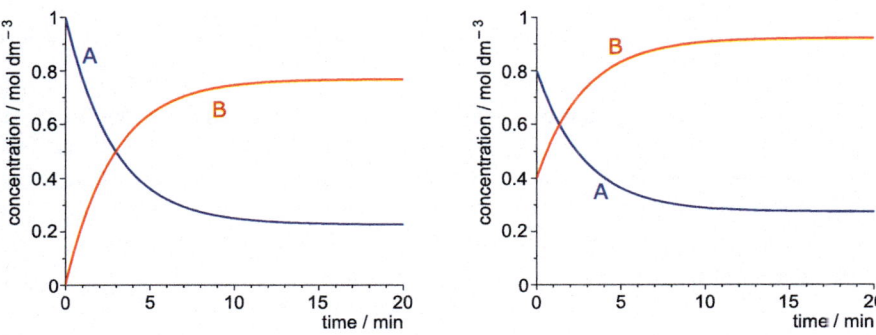

Fig. 4.3 Concentration profile of the reversible reaction $A \underset{k_{-1}}{\overset{k_1}{\rightleftharpoons}} B$. The left panel shows the evolution of the concentrations in case when there is only component A and no B in the system at $t = 0$. The right panel shows the case when the reaction mixture contains also some of the component B at $t = 0$ in addition to component A. During the course of reaction, the stoichiometric relation $[A] + [B] = [A]_o + [B]_o$ always holds. After long enough time ($t = t_\infty$) for the equilibrium to develop, the ratio of concentrations is given by the equilibrium constant $= \frac{[B]_\infty}{[A]_\infty} = \frac{k_1}{k_{-1}}$

can always write $[A] + [B] = [A]_o + [B]_o$ and express the concentration of component B as $[B] = [A]_o + [B]_o - [A]$. Thus, the only complication is that we should write $[A]_o + [B]_o$ in place of the constant $[A]_o$ in the previous case. Consequently, the rate equation according to the new initial condition reads as

$$\frac{d[A]}{dt} = -k_1[A] + k_{-1}\left([A]_o + [B]_o - [A]\right). \tag{4.47}$$

In analogy to the result in the previous case, we can readily write the particular solution as,

$$[A] = \left([A]_o + [B]_o\right)\frac{k_{-1}}{k_1 + k_{-1}} + [A]_o e^{-(k_1+k_{-1})t} - \left([A]_o + [B]_o\right)$$
$$\times \frac{k_{-1}}{k_1 + k_{-1}} e^{-(k_1+k_{-1})t}. \tag{4.48}$$

Using similar rearrangements as before, we can obtain a more compact form as well:

$$[A] = \frac{k_{-1}}{k_1 + k_{-1}}\left([A]_o + [B]_o\right) + \frac{k_1[A]_o - k_{-1}[B]_o}{k_1 + k_{-1}} e^{-(k_1+k_{-1})t}. \tag{4.49}$$

We can see that, if we write zero in place of $[B]_o$ in this expression, we get back the solution of the previous case. The concentration of component B can also be calculated similarly to the previous case, making use of the stoichiometry:

$$[B] = \frac{k_1}{k_1 + k_{-1}} \left([A]_o + [B]_o\right) - \frac{k_1[A]_o - k_{-1}[B]_o}{k_1 + k_{-1}} e^{-(k_1 + k_{-1})t}. \tag{4.50}$$

It is worth exploring another interesting property of this solution. Being a reversible reaction, after long enough time ($t = t_\infty$) the rate of the opposing reactions will be the same and equilibrium will be reached. The condition of the equilibrium can be written as $k_1[A]_\infty = k_{-1}[B]_\infty$, from which we can express $[B]_\infty = [A]_\infty \frac{k_1}{k_{-1}}$. Let us substitute this expression in the identity $[A]_o + [B]_o = [A]_\infty + [B]_\infty$. The constant term in the above solution is simplified this way:

$$\frac{k_{-1}}{k_1 + k_{-1}} \left([A]_o + [B]_o\right) = \frac{k_{-1}}{k_1 + k_{-1}} [A]_\infty \left(1 + \frac{k_1}{k_{-1}}\right)$$

$$= \frac{k_{-1}}{k_1 + k_{-1}} [A]_\infty \left(\frac{k_1 + k_{-1}}{k_{-1}}\right) = [A]_\infty. \tag{4.51}$$

Substituting this into the expression of the concentration of component A, we get

$$[A] = \frac{k_1[A]_o - k_{-1}[B]_o}{k_1 + k_{-1}} e^{-(k_1 + k_{-1})t} + [A]_\infty. \tag{4.52}$$

Similarly, we can write the concentration of component B the following way:

$$[B] = \frac{k_{-1}[B]_o - k_1[A]_o}{k_1 + k_{-1}} e^{-(k_1 + k_{-1})t} + [B]_\infty. \tag{4.53}$$

The equilibrium concentration after it has been stabilised is easy to determine, unlike the initial concentrations that change in time until there is equilibrium; thus, these expressions are of practical importance.

4.5 Solving Rate Equations of Mechanisms Comprising First-Order Reactions Only

For reaction mechanisms comprising only first-order elementary reactions (or mechanistic steps) always exist closed-form solutions which are linear combinations of exponential functions. In case of complicated mechanisms, this solution is usually calculated using one of two methods that we shall discuss here.

First, we shall discuss the method based on the use of operational calculus. According to this, linear differential equations with constant coefficients can be transformed into a linear algebraic equation using *Laplace transforms*. We shall not discuss details of operational calculus here, only the principles of solving differential equations using Laplace transforms. The Laplace transform of a time-dependent

4.5 Solving Rate Equations of Mechanisms Comprising First-Order Reactions Only

function $c(t)$ is an integral which depends on another variable s and is calculated the following way:

$$C(s) = \int_0^\infty e^{-st} c(t) dt. \qquad (4.54)$$

As a result of this transformation, we get a linear algebraic equation containing the transform $C(s)$. Performing the transformation for the rate equations of each component, we get a system of linear algebraic equations whose solution gives the functions $C_i(s)$. Applying then an inverse Laplace transformation of these functions and considering initial conditions we get the solutions $c_i(t)$ for the concentration of the components (the inverse Laplace transform is somewhat more complicated, also containing complex numbers).

Calculating Laplace- and inverse Laplace transforms is considerably cumbersome. Instead of calculating them, previously calculated transforms and inverse transforms of many functions had been tabulated. However, at the beginning of the twenty-first century, software packages developed for symbolic computations are available which can readily perform these transformations and also solve systems of linear algebraic equations, by issuing simple commands in a user-friendly manner. Well known examples are Mathematica, MathCad or Maple. With their help, in case of moderately complicated mechanisms, we can easily get solutions for the rate equations.

The other method is based on matrix-vector calculus. The underlying principle of this method is the fact that solutions for mechanisms containing first-order steps only are always linear combinations of exponential functions. We shall illustrate the method on the example of the simple consecutive reaction discussed earlier in Sect. 4.3:

$$A \xrightarrow{k_1} B \xrightarrow{k_2} C. \qquad (4.55)$$

Let us write the rate equations for the three components in the following forms:

$$\begin{aligned}
\frac{d[A]}{dt} &= -k_1[A] + 0[B] + 0[C] \\
\frac{d[B]}{dt} &= +k_1[A] - k_2[B] + 0[C] \\
\frac{d[C]}{dt} &= +0[A] + k_2[B] + 0[C]
\end{aligned} \qquad (4.56)$$

We can see that this system of differential equations can readily be written using matrix-vector notation,

$$\left(\frac{d[A]}{dt}, \frac{d[B]}{dt}, \frac{d[C]}{dt}\right) = \begin{pmatrix} -k_1 & 0 & 0 \\ +k_1 & -k_2 & 0 \\ 0 & -k_2 & 0 \end{pmatrix} \begin{pmatrix} [A] \\ [B] \\ [C] \end{pmatrix}. \tag{4.57}$$

The left-hand side is a row vector containing time-derivatives of the concentrations, which is identical to the time-derivative of the concentration vector. The first factor on the right-hand side is the so-called *kinetic matrix* or *rate-coefficient matrix*. The column vector of the concentrations is multiplied by this matrix. Thus, we can rewrite this equation in a simplified notation using the vector $\underline{c}$ of n components and the $n \times n$ matrix $\boldsymbol{K}$:

$$\frac{d}{dt}\underline{c} = \boldsymbol{K}\,\underline{c} \text{ or } \dot{\underline{c}} = \boldsymbol{K}\,\underline{c}, \tag{4.58}$$

using the shorthand notation $\dot{\underline{c}}$ for the time-derivative of the vector $\underline{c}$.

Let us make use now of the fact that the solution of any mechanism consisting only of first-order steps is a linear combination of exponential functions,

$$\underline{c} = \sum_i a_i e^{\lambda_i t}. \tag{4.59}$$

The time-derivative of this sum of exponential functions is readily written as

$$\frac{d}{dt}\underline{c} \equiv \dot{\underline{c}} = \sum_i \lambda_i a_i e^{\lambda_i t}. \tag{4.60}$$

Vector $\underline{c}$ is the *particular* solution of the system of differential equations (4.57), which can be written as the linear combination of the general solutions $\underline{x}_i$ of the same system of differential equations: $\underline{x}_i = \sum_j r_j e^{\lambda_i t}$; for the case of λ_i in the exponent.

As $\underline{x}_i$ is the (general) solution of the system of differential equations, we can write the equation containing the kinetic matrix $\boldsymbol{K}$ also as

$$\dot{\underline{x}}_i = \boldsymbol{K}\underline{x}_i. \tag{4.61}$$

As $\underline{x}_i$ is an exponential function, its derivative with respect to t can be written as $\dot{\underline{x}}_i = \lambda_i \underline{x}_i$. Consequently, instead of the original system of differential equations, we have a simple system of *algebraic* equations,

$$\boldsymbol{K}\underline{x}_i = \lambda_i \underline{x}_i. \tag{4.62}$$

This is the *eigenvalue equation* of the matrix $\boldsymbol{K}$, where the vector $\underline{x}_i$ is the *eigenvector* corresponding to the eigenvalue λ_i.

4.5 Solving Rate Equations of Mechanisms Comprising First-Order Reactions Only

We can sum up the solution of the original system of differential equations $\underline{\dot{c}} = \boldsymbol{K}\,\underline{c}$ the following way: solve the eigenvalue equation $\boldsymbol{K}\underline{x}_i = \lambda_i \underline{x}_i$, then choose the linear combination coefficients a_i of the obtained eigenvectors so that they satisfy the initial conditions.

Again, we shall not discuss details of matrix calculus here, only the principles to solve the above equation to get the vector $\underline{c}$ as a function of time. To find eigenvalues and eigenvectors of the matrix, we should solve the following equation (which is called the *characteristic equation* of matrix $\boldsymbol{K}$);

$$(\boldsymbol{K} - \lambda \boldsymbol{I})\,\underline{r} = 0. \tag{4.63}$$

Here, λ is an eigenvalue, $\underline{r}$ is an eigenvector and $\boldsymbol{I}$ is the unit matrix whose diagonal elements are units and all off-diagonal elements are zeros. For each of the n eigenvalues λ_i, there exists an eigenvector $\underline{r}_i$. Eigenvectors multiplied by any constant number are also eigenvectors of the matrix. Thus, we are looking for the general solution of the system of differential equations in the form of

$$\underline{x}_i = \underline{r}_i e^{\lambda_i t}, \tag{4.64}$$

as we know that the solutions should be linear combinations of exponential functions. Any linear combination of the eigenvectors also satisfies the eigenvalue–eigenvector equation (4.63). Accordingly, particular solutions of the system of differential equations can be obtained if these linear combinations also satisfy the initial conditions $\underline{c}$ ($t = 0$). Substituting initial conditions into the system of equations (4.64) and solving them, we get the searched for linear combination coefficients.

Calculations outlined here are quite elaborate, but software packages for symbolic computations mentioned above can readily make perform these calculations by issuing simple commands in a user-friendly manner. It is enough to specify the kinetic matrix $\boldsymbol{K}$ and the concentration vector $\underline{c}$ containing the initial conditions at $t = 0$ to get a solution.

The two methods discussed are really efficient to solve the system of differential equations. However, if reaction mechanisms are too much complicated, they do not provide explicit results using symbolic computation software packages, or, in many cases, the solution might be overly complicated. In case of mechanisms that contain other than first-order steps, even those that are only slightly complicated, would not have analytical solutions. In these cases, it is more appropriate to find solutions using numerical instead of analytical methods. Such methods to find solutions are discussed in a later section of this chapter.

4.6 Quasi-Steady-State Approximation

The more complicated a system of differential equations, the more effort is needed to solve it. If we can simplify the rate equations in some way, their solution will be easier as well. The historically first simplification method – the *quasi-steady-state approximation* or, using the common acronym, QSSA[1] – has been implemented more than a century before. This method uses the approximation that the concentration of some species – which are highly reactive and are usually present only in a small concentration – can be considered as constant during the course of reaction (typical examples are reactive radicals).

We shall discuss this method for the case of an overall reaction that frequently occurs in the chemical praxis. The reaction can be written as follows:

$$A + B + D = E. \tag{4.65}$$

Component C has been omitted in this equation as there is an intermediate in the mechanism that is produced from components A and B in a reversible reaction, which then gives the product E when reacting with component D. This mechanism can be written as

$$A + B \underset{k_{-1}}{\overset{k_1}{\rightleftarrows}} C \tag{4.66}$$

$$C + D \xrightarrow{k_2} E. \tag{4.67}$$

Let us write the rate equations first, according to the rules that have been discussed earlier.

$$\begin{aligned} \frac{d[A]}{dt} &= -k_1[A][B] + k_1[C] \\ \frac{d[B]}{dt} &= -k_1[A][B] + k_1[C] \\ \frac{d[C]}{dt} &= k_1[A][B] - k_{-1}[C] - k_2[C][D] \\ \frac{d[D]}{dt} &= -k_1[C][D] \\ \frac{d[E]}{dt} &= k_2[C][D] \end{aligned} \tag{4.68}$$

---

[1]This method has been first suggested by the German chemist Max Ernst August Bodenstein (1871–1942) in 1913, supposing approximate stationary concentrations of some intermediate species during the course of reaction. Accordingly, the original name was given in German: "Quasistationaritätsprinzip".

4.6 Quasi-Steady-State Approximation

Let us apply the QSSA by considering the concentration of the intermediate C as a time-independent constant. It is obvious that, at the beginning of the reaction, when only reactants A, B and D are present, this cannot be true; only after some time t'. The quasi-constant concentration develops due to the fact that component C is formed only slowly but reacts rather quickly, and, after the time t' its *rate of decomposition will be* (approximately) *the same as its rate of formation*,

$$k_1[A][B] \cong k_{-1}[C] + k_2[C][D]. \tag{4.69}$$

This equation is called the *steady-state hypothesis*.

The consequence of this hypothesis is that we can change the differential equation for $\frac{d[C]}{dt}$ in the mechanism (4.68) to the algebraic equation (4.69), as it contains all information which is contained in the original differential equation in case of the validity of the hypothesis. The result is a simpler, easier-to-solve *coupled system of differential-algebraic equations*. In the current example, we can express the concentration of component C from equation (4.69) as a function of the concentrations of components A, B and D,

$$[C] = \frac{k_1[A][B]}{k_{-1} + k_2[D]}. \tag{4.70}$$

Upon substitution of this expression in place of [C] in the remaining differential equations, we eliminate the concentration of component C from them; thus, their solution will be easier. If we are interested only in the rate of the formation of the product E, we can express it in a compact form that does not contain the concentration of the intermediate C, by substituting the above expression into the last differential equation of the system (4.68),:

$$\frac{d[E]}{dt} = k_2 \frac{k_1[A][B]}{k_{-1} + k_2[D]}[D] = \frac{k_1[A][B]}{1 + k_{-1}/k_2[D]}. \tag{4.71}$$

An important consequence of applying the QSSA is what we can see in this expression; when measuring the rate of formation of the product E as a function of the concentrations of reactants A, B and D, we can determine the rate coefficient k_1 but neither k_{-1} nor k_2, only their ratio. The latter two could only be determined if we could measure the concentration of C as a function of time, and would not apply the QSSA (Fig. 4.4).

Another interesting property of the approximation reveals the fact that it is only an approximation. The time-derivative of the expression contained in equation (4.70) clearly shows that it cannot be zero; that is, the concentration of the QSS component C cannot be strictly constant,

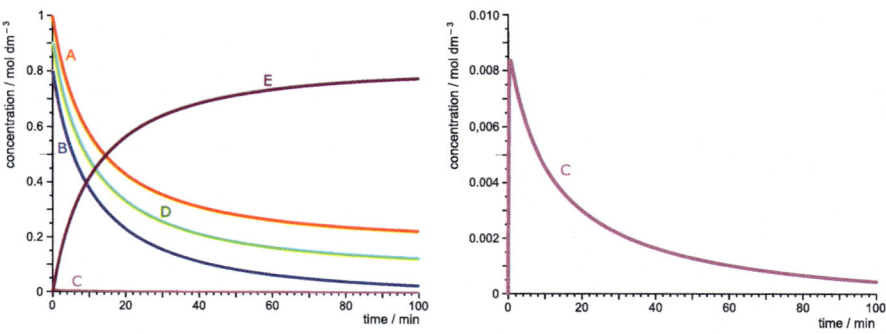

Fig. 4.4 Applicability of QSSA for the composite reaction comprising steps $A + B \underset{k_{-1}}{\overset{k_1}{\rightleftharpoons}} C$ and $C + D \xrightarrow{k_2} E$. The left panel shows the evolution of the concentrations of all reacting species. Note that the concentrations of the reactants and the products change a lot compared to the concentration of the intermediate C. The right panel shows the temporal evolution of the concentration of the intermediate C at a 100 times enlarged scale. It is readily seen in this panel that the concentration of the quasi-steady-state species is not at all constant as a function of time. However, in order to provide the precursor for the fast formation of the end product E, its formation and decomposition should be fast enough – practically the same as that of the other components. This is the underlying origin of its low concentration: there is very little difference between the rate of its formation and decomposition, thus, the steady-state hypothesis (4.69) can be applied for its concentration. Equivalent formulation of the hypothesis is that the concentration of the intermediate C is strongly coupled to the (non-steady-state) concentration of the reactants and products. Reaction rate coefficients are $k_1 = 0.1$ dm^3/(mol min), $k_{-1} = 0.01$ min^{-1} and $k_2 = 10$ dm^3/(mol min)

$$\frac{d[C]}{dt} = \frac{d}{dt}\left(\frac{k_1[A][B]}{1 + k_{-1}/k_2[D]}\right). \tag{4.72}$$

Summing up we can state that the underlying basis of the QSSA is the negligible difference in the rates of formation and decomposition reactions of the QSS component. Each QSS component results in changing the relevant differential equation into an algebraic one, thus simplifying the solution of the system of the original differential equations. There are reactions in which we can hypothesise several QSS components, if the condition to do so holds. In most of these cases, the solution of the underlying system of algebraic equations provides the steady-state concentration of the QSS species. Substituting these into the remaining differential equations, their solution becomes simpler. Generally, the rate equation of the final product will also become simpler.

4.7 Fast Pre-equilibrium Approximation

Let us consider again the mechanism given by equations (4.66) and (4.67), using different conditions and a different approximation. When the rate of the first two steps is much greater than that of the third step, we expect the equilibrium (4.66) to

4.7 Fast Pre-equilibrium Approximation

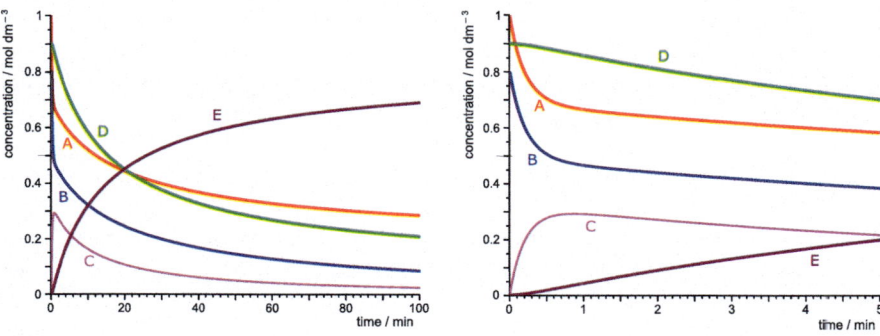

Fig. 4.5 Applicability of the approximation of fast pre-equilibrium for the composite reaction comprising steps $A + B \underset{k_{-1}}{\overset{k_1}{\rightleftarrows}} C$ and $C + D \overset{k_2}{\longrightarrow} E$. In the left panel, we can see that equilibrium with respect to species A, B and C is reached quite fast, concentrations of the species will then fulfil the condition (4.73). The right panel shows – with a 20-times enlarged time scale – that the equilibrium is reached within 1 minute, while the reaction takes substantially more than 100 minutes to complete. Reaction rate coefficients are $k_1 = 2$ dm^3/(mol min), $k_{-1} = 2$ min^{-1} and $k_2 = 0.2$ dm^3/(mol min)

be reached very fast and maintained during the entire course of reaction. From the condition of the equilibrium $A + B \rightleftharpoons C$,

$$K = \frac{[C]}{[A][B]}, \qquad (4.73)$$

we can express the concentration of the intermediate C at any instance of time as $[C] = K[A][B]$. Upon substitution of this expression into the last rate equation of the mechanism (4.68), we can obtain the rate of the formation of product E in a simple form,

$$\frac{d[E]}{dt} = k_2[C][D] = k_2 K\,[A][B][D]. \qquad (4.74)$$

As we can see, this is formally a third-order reaction rate with the rate coefficient $k_2 K$. It is also clear that we can determine the rate coefficient k_2 from kinetic measurements only if we know the equilibrium constant K from an independent experiment. From kinetic measurements of the reaction, we can only determine the product $k_2 K$.

Note that QSSA is not equivalent to the fast pre-equilibrium approximation. In this example, the QSS hypothesis holds if the approximation (4.69) is valid (within the allowed limit of error); however, this does not guarantee that the equilibrium concerning reaction (4.66) also holds. As all the concentrations [A], [B] and [D] are time dependent, the concentration change rate of C depending on them will not be zero – which seemingly contradicts the QSS hypothesis. This is one of the reasons that we have formulated the QSS hypothesis (4.69) in the form given there. There is also another reason to present the equation in this form; namely, the reason concerning the limit of error allowed in the approximation. Many texts state the QSSA hypothesis in the form which is (almost) equivalent to (4.69) in this particular example as

$$k_1[A][B] - k_{-1}[C] - k_2[C][D] \cong 0. \tag{4.75}$$

It is easy to recognise that this form is misleading. The approximation in the form (4.69) can be interpreted the way that the difference of the rate at the left-hand side and the rate at the right-hand side can be neglected with respect to the rates themselves; in the case of the approximation written as equation (4.75), the validity of the difference of the left-hand side from zero cannot be judged to hold, as zero itself is infinitesimally small. Accordingly, we shall always formulate the QSS hypothesis in this book by the approximate equality of the rate of formation and the rate of decomposition of the QSS components. We shall exploit the usefulness of QSS hypothesis later, on the example of chain reactions; particularly, the formation of HBr from their elements.

One of the first examples discovered for a fast pre-equilibrium approximation as an alternative for a termolecular reaction was the oxidation of nitrogen monoxide to nitrogen dioxide. The overall equation of the reaction is the following:

$$2NO + O_2 \rightarrow 2NO_2. \tag{4.76}$$

Kinetic experiments supported that the reaction is of third order, that is, the rate of formation of the product can be written as

$$\frac{d[NO_2]}{dt} = k[NO]^2[O_2]. \tag{4.77}$$

There is also experimental evidence that the rate of reaction *diminishes* with increasing temperature, though the reaction is not very fast. From this we could conclude – based on the Arrhenius equation as well as on the extended Arrhenius equation – that the activation energy of the reaction is negative. However, elementary reactions cannot have negative activation energy, and activation energy close to zero implicates a very fast reaction whose rate is close to the collision frequency of the molecules. Consequently, this reaction should be a composite one, and we should also account for the unusual temperature dependence. Both expectations are fulfilled by the following mechanism:

$$2\,NO + O_2 \underset{k_{-1}}{\overset{k_1}{\rightleftharpoons}} N_2O_2 \tag{4.78}$$

$$N_2O_2 + O_2 \xrightarrow{k_2} 2\,NO_2. \tag{4.79}$$

Let us apply first the hypothesis of fast pre-equilibrium, supposing that $k_1 \gg k_2$ and $k_{-1} \gg k_2\,[O_2]$. Under these conditions, we can use the equilibrium constant

$$K = \frac{[N_2O_2]}{[NO_2]^2} \tag{4.80}$$

to get the time-dependent concentration of the intermediate N_2O_2,

$$[N_2O_2] = K[NO_2]^2. \tag{4.81}$$

Let us substitute this into the rate equation of formation of NO_2 according to the bimolecular reaction (4.79),

$$\frac{d[NO_2]}{dt} = k_2[N_2O_2][O_2] = k_2K[NO]^2[O_2]. \tag{4.82}$$

Here we can see that this result is identical to the experimental rate (4.77), with the additional information that the rate k in the equation is the product k_2K. This also helps to interpret the unusual temperature dependence. The reaction of pre-equilibrium (4.78) is an exothermal one, whose equilibrium constant diminishes with increasing temperature. The rate coefficient k_2 of the reaction (4.79) – having a relatively low activation energy – increases more slowly with increasing temperature than the equilibrium constant K decreases. As a result, the product k_2K decreases with increasing temperature, (falsely) suggesting a negative activation energy (it is generally true that negative apparent activation energy indicates a complex reaction mechanism).

4.8 Rate-Determining Steps

In the previous section where we use the approximation of fast pre-equilibrium, we can also conclude that the rate-determining step in the mechanism (4.66)–(4.67) – according to the pre-equilibrium hypothesis – is the third reaction (4.67), as the rate equation describing the formation of product E contains only the rate coefficient k_2 referring to this reaction.

This conclusion can also be stated in a more general case, considering the QSSA result (4.71), without invoking the fast pre-equilibrium approximation. Namely, if in the denominator of this expression the condition $k_{-1} \gg k_2[D]$ (*one* of the conditions only of the applicability of fast pre-equilibrium approximation) holds, then the term $k_2\,[D]$ can be neglected compared to k_{-1}, and we can obtain the same expression,

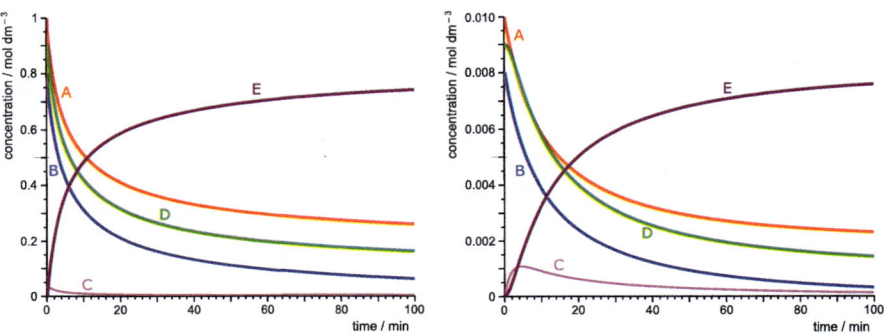

Fig. 4.6 Illustration of different rate-determining steps in case of the mechanism comprising the three reactions $A + B \underset{k_{-1}}{\overset{k_1}{\rightleftharpoons}} C$ és $C + D \overset{k_2}{\rightarrow} E$. Left panel: $k_1 = 10$ dm^3/(mol min), $k_{-1} = 200$ min^{-1} and $k_2 = 5$ dm^3/(mol min). Here, the condition $k_{-1} \gg k_2 [D]$ holds and the rate-determining step is the formation of the product E with a rate law of third order (note that the condition of pre-equilibrium does not hold as k_1 and k_2 are close to each other). Right panel: $k_1 = 0.1$ dm^3/(mol min), $k_{-1} = 0.01$ min^{-1} and $k_2 = 0.5$ dm^3/(mol min). Accordingly, the condition $k_{-1} \ll k_2[D]$ holds here, and the rate-determining step is the formation of the intermediate C with a rate law of second order (the condition of pre-equilibrium does not hold here either as k_1 and k_2 are not much different)

$$\frac{d[E]}{dt} = \frac{k_1 k_2 [A][B][D]}{k_{-1} + k_2[D]} \cong k_1 \frac{k_1}{k_{-1}} [A][B][D] = k_2 K \, [A][B][D]. \quad (4.83)$$

If we have a reverse relation $k_{-1} \ll k_2[D]$, the term k_{-1} can be neglected compared to $k_2 [D]$

$$\frac{d[E]}{dt} = \frac{k_1 k_2 [A][B][D]}{k_{-1} + k_2[D]} \cong \frac{k_1 k_2 [A][B][D]}{k_2[D]} = k_1 \, [A][B], \quad (4.84)$$

which means that the rate-determining step within these conditions is the formation of the intermediate C and the reaction (applying the QSS hypothesis for the component C) obeys second-order kinetics. In this case, we can determine only the rate coefficient k_1 from kinetic measurements.

Based on the examples shown above, we can draw an important conclusion concerning the results of kinetic experiments and the mechanism constructed based on these results. We have seen that, under some specific conditions, experimental results can be interpreted with a second-order rate law, while under other conditions, with a third-order rate law. As we have seen, we cannot derive from this result that the reaction would be an elementary one of either second or third order. In case of a second-order rate law – which does not contain the concentration of the species D – it is evident that the reaction should have a complex mechanism containing steps involving also component D. In the case of the third-order result, there is no such evident hint in favour of a composite reaction. However, if the reaction is fast enough, we could hypothesise that it cannot be a termolecular (elementary) reaction, as its probability and its rate is quite low, especially in gas phase. Thus, if we suspect a composite reaction, we need further experiments exploring other conditions and looking for additional intermediate species. As we have seen in connection with the example of the reaction of NO with O_2 in Sect. 4.7, thermodynamic considerations can also help to find relevant alternative composite reactions.

We can also draw the general conclusion that we cannot 'prove' the validity of a reaction mechanism; all we can state is that the mechanism suggested can account for all available experimental observations. In this case, we can accept the mechanism as a valid model of the reaction under the conditions of the experimental observations. However, it is not surprising that an accepted mechanism for a reaction often needs a reformulation as a more complicated mechanism after more detailed experimental observations are available.

4.9 Numerical Algorithms to Solve Differential Equations; Integrators for Reaction Kinetics

As it is mentioned in Sect. 4.3 when solving differential equations for a third-order reaction, an analytical (closed form) solution of the rate equations does not always exist, not even in an implicit form which could be expressed with elementary functions. For composite reactions, this is the typical case. Even if we can find an analytical form, in many cases it is so much complicated (just to write it down would take several pages) that it is not practical to use for actual calculations. As this is a frequent case, we need the solution of differential equations anyway to be able to calculate time-dependent concentrations during the course of reaction. In these cases, we compute the numerical approximation of the solution using relevant numerical methods.

We shall briefly describe this procedure here, using a simple example which is easy to extend for more complicated cases as well. Let us have a rate equation in the following form:

$$\frac{dc(t)}{dt} = f(c(t)) \qquad (4.85)$$

(the derivative $f(c)$ on the right-hand side is also a function of time, as the concentration c itself is a function of time. However, to simplify notation, it is usually not written explicitly). We can separate the above differential equation to get:

$$dc(t) = f(c(t))dt. \qquad (4.86)$$

Accordingly, we can write the actual concentration at time $t + dt$ as

$$c(t + dt) = c(t) + f(c(t))dt. \tag{4.87}$$

If we would proceed in time with infinitesimally small steps, we would never arrive to the end of a finite time interval; therefore, we should rather calculate a later concentration using a small but finite increment of time Δt and the derivative $f(c(t))$. Due to this finite increment, a definite integral appears in the expression,

$$c(t_i + \Delta t) = c(t_i) + \int_{t_i}^{t_i + \Delta t} f(c(t))dt. \tag{4.88}$$

If the increment Δt is small enough, we can neglect the change of the derivative within this time-step (concentration functions during the course of reactions are smooth and 'well-behaved'; thus we can expect this relatively small change). Taking this property into consideration, we can replace the integral with a simple product as follows:

$$c(t_i + \Delta t) = c(t_i) + f(c(t_i))\Delta t. \tag{4.89}$$

As we can see, the numerical integration can be done step by step. Thus, after the n-th increment, the $n + 1$-th value of the concentration function can be calculated the following way:

$$c_{n+1} = c_n + f(c_n) \cdot (t_{n+1} - t_n). \tag{4.90}$$

To perform the integration stepwise, we should start from a known concentration value; but this is not a problem as we usually know the initial concentration c_o at the start of the reaction (this is the reason to call this method as an *initial value problem*). Another problem is to find the value of the derivative function $f(c_n)$ at the concentration c_n.[2]

To sum up, we can state that the initial value problem

$$\frac{dc(t)}{dt} = f(c(t)); c(t = 0) = c_o, \tag{4.91}$$

can be solved using the algorithm outlined above; the smaller the time-step Δt (with special attention to the slope of the function), the more precisely.

---

[2]Numerical integration methods *to solve differential equations* are different from the numerical integration methods of *known functions* (e.g. trapezoidal rule, Simpson's rule or Gaussian quadrature formulas). In the former case, we only know the solution function $c(t_i)$ in discrete instances of time; thus, we should also approximate the values of the derivative function numerically at each instance of time.

When implementing this algorithm, special attention should be paid to several things. The time-step Δt should evidently be smaller than the time when a non-negligible change of the derivative would occur (in case of a roughly linear function, it can be longer, but in case of sudden changes, it should be short enough). Therefore, there is also a necessity to calculate not only the derivative during the stepwise integration but also the optimal step-length. To get a good approximation for the derivative also in case of steeper concentration change, it is not calculated simply by dividing the increment of the function between adjacent values by the time-step, but by fitting a polynomial to several measured (or sometimes interpolated) values and calculating its derivative at the desired instance of time. The next concentration will then be calculated using this derivative in the recursive formula (4.90).

Another problem arising during numerical integration is the accumulation of the error in subsequent steps. The smaller the time-step Δt, the smaller the error of the derivative and that of the increment of the concentration in one integration step. However, there is always some *round-off error* due to the finite number of digits in the calculations and a so-called *truncation error* due to the truncation of the polynomial to calculate the derivative. These errors will accumulate during the integration steps; thus, it is not practical to increase too much the number of steps (which means decreasing the step size). The optimal step size is mostly determined in an adaptive way.

The simplest way to adaptively determine the step-length is *halving the step size*, which provides information concerning the error of the numerical approximation. Using this method, first the derivative $f(c)$ and the concentration $c(t)$ is calculated using the step size Δt. The same calculations will then be performed using two steps with $\Delta t/2$ step size. Then the first $c(t)$ value is compared to the one calculated after two steps of size $\Delta t/2$ only. The difference of the two values provides the error of not halving the original step size. If this error is greater than a pre-selected *tolerance*, than either $\Delta t/2$ is used as step size, or a suitable algorithm is applied to calculate the optimal step size as a function of the error and the tolerance (an extra benefit of the more complicated calculation is that the next concentration value is always calculated from the more accurate two steps; even if step size Δt should not be halved).

In reaction kinetics, the most popular numerical integration method is the one called Runge–Kutta[3] with adaptive step size. If there is no further specification of this method, then it refers to the *fourth-order* Runge–Kutta method which calculates the approximation of the derivative up to the fourth-order term of the polynomial (its usual acronym is RK4).

When considering the use of freely available numerical integrator applications for reaction kinetic purposes at the end of the 2010s, we should check for their following

---

[3]Carl David Tolmé Runge (German mathematician; 1856–1927) has improved in 1901 the method of Martin Wilhelm Kutta (another German mathematician; 1867–1944) conceived in 1895 to numerically solve systems of ordinary differential equations. Adaptive step-length calculations have been developed in the 1960s by several mathematicians.

properties. Some of them are developed to solve the rate equations of homogeneous composite reactions taking place in a single 'container'. Some others can easily treat homogeneous composite reactions taking place in different 'compartments' with the possibility of material transport between them (in the latter case, functions describing component transfer should also be specified. Typical applications of these software packages are biochemical reactions within living cells, in the extracellular matrix, and between these compartments). Input data for these integrators are typically stoichiometric equations of elementary (or mechanistic) steps of composite reactions, functions describing material transport between compartments and parameters involved in them (*e.g.* rate coefficients, initial concentrations and flow rates). Applications typically transform stoichiometric equations into rate equations (if they are defined as mass-action rate equations), but there is also possibility to give user-defined functions describing time-dependent concentrations.

Reaction kinetic integrator applications are capable of performing plenty of calculations which necessitate numerical integration. Most important options are simulations of concentration versus time functions for the reacting components involved in the given mechanism, and its relevant parameters, and estimation of kinetic parameters (*e.g.* rate coefficients, initial concentrations and equilibrium constants) along with their standard deviations and other important statistical properties (*e.g.* covariance matrix, information matrix and sensitivity coefficients). Before starting the calculations, the user can specify what kind of integrator algorithm should be used and what the expected tolerance is. The user can also specify the content of the output, can ask for data storage after specified time periods, and can have access to a variety of graphically displayed information.

There is also an important option offered by modern chemical integrators: the use of so-called *stochastic kinetics*. Without going into details of this topic, it is worth to know that if individual molecules do play some role (*i.e.* rate equations referring to average bulk concentrations do not give satisfactory results), stochastic kinetic methods calculate the progress of the reaction based on the probability of reaction of individual molecules. This option is important for example when there are only a few or at highest a few hundred molecules of certain species in a compartment (intracellular biochemical reactions often have this complication). At the start of calculations, the user usually can choose between deterministic or stochastic methods. In the latter case, calculations are performed not using concentration of the components but their number of molecules.

Kinetic integrator applications are developing quite quickly, partly due to the development of numerical methods, partly to an increasing number of tasks they can automatically do without effort from the user. In the light of this development, there is no use to cite here actual software packages; it is left to the reader to find the suitable one available.

It is worth to remind the reader at the end of this section that rate equations of composite reactions comprise a system of ordinary differential equations. Numerical solution of these equations is quite similar to the single variable problem to solve one single differential equation. The only difference is that, for each component at the actual concentration values $c_j(t_i)$, the derivative according to the time step Δt should be computed, then after an increment Δt, the next concentration values $c_j(t_i + \Delta t)$

should be computed via numerical integration. A simple practical formulation of this algorithm is that we build a vector $\underline{c}(t_i)$ containing the concentrations $c_j(t_i)$, then calculate the derivative of this vector for the subsequent numerical integration step. Integrator applications for reaction kinetic purposes usually perform these operations automatically; thus, even in the case of a complicated composite mechanism, it is enough to give the stoichiometric equations of the steps of the actual reaction mechanism by the user.

4.10 Chain Reactions

Previous sections describe in detail possible connections of elementary reactions. However, an interesting coupling of reactions was not discussed in which several reactions are coupled in a way that they form a closed loop that can proceed in round repeatedly until all the necessary reactive components are available in the reaction mixture (see Scheme (4.99)). This type of reaction mechanism is called *chain reaction*. An important property of chain reactions is that, once a highly reactive key species – typically a radical or an atom – is formed, it drives the reaction loop to proceed in the presence of the reactants lots of times, always regenerating this highly reactive species. These reactions can be divided into two groups: *unbranched* chain reactions and *branched* chain reactions. This section describes them in more detail.

4.10.1 Unbranched Chain Reactions

Here we illustrate this type of reaction by discussing one of the first studied examples, the formation of hydrogen-bromide from its elements. Immediately after mixing gas phase H_2 and Br_2, a slow reaction begins at moderately higher than room temperature, whose initial rate can be given by the following differential equation:

$$\frac{d[\text{HBr}]}{dt} = k[\text{H}_2][\text{Br}_2]^{1/2}. \tag{4.92}$$

However, when the reaction proceeds, this rate equation is less and less valid, and the rate of formation of the product can be described by the following differential equation:

$$\frac{d[\text{HBr}]}{dt} = k\frac{[\text{H}_2][\text{Br}_2]^{1/2}}{1 + k'\frac{[\text{HBr}]}{[\text{Br}_2]}}. \tag{4.93}$$

(Note that we get back the previous rate equation by substituting zero in place of the concentration of HBr). One striking feature is the non-integer exponent ½ in the numerator; the other surprising feature is that the right-hand side of equation (4.93) has not the familiar form of a homogeneous n-th order expression. Obviously, we cannot assign any 'reaction order' to such an expression (most composite reactions have this property; the notion of reaction order is not applicable for them). Equation (4.93) has been verified experimentally in 1907 but the underlying kinetic principle was only discovered in 1919, 12 years after the detailed experiments. (Since that time this is the favoured textbook example for the chain reaction. As it turned out later, a fractional exponent in the rate equation often indicates a chain reaction).

The overall reaction $H_2 + Br_2 \rightarrow 2\,HBr$ has the following mechanism:

$$Br_2 + M \xrightarrow{k_1} 2Br + M \quad \text{initiation} \tag{4.94}$$

$$Br + H_2 \xrightarrow{k_2} HBr + H \quad \text{propagation} \tag{4.95}$$

$$H + Br_2 \xrightarrow{k_3} HBr + Br \quad \text{propagation} \tag{4.96}$$

$$H + HBr \xrightarrow{k_4} H_2 + Br \quad \text{inhibition} \tag{4.97}$$

$$2Br + M \xrightarrow{k_5} Br_2 + M \quad \text{termination} \tag{4.98}$$

The symbol M in the reactions (4.94) and (4.98) may be a molecule of any species. Its role in the initiation reaction is to transfer sufficient energy to a Br_2 molecule when colliding with it; while in the termination step it is the 'third body' that absorbs the energy of the recombination – which is an exothermal reaction – from the newly formed Br_2 molecule to prevent its re-dissociation (if there are also inert (non-reacting) species in the reaction mixture, their molecules can also play the same roles).

The loop of reactions can be better visualised by drawing the directed multigraph of the reaction, the way biochemical reaction networks are often depicted. In this scheme, reactions are represented by a long (in case of more than one product: bifurcating) arrow. Reactants are at the tail, products at the head of the arrow. In case of more than one reactant, reactants can be written at the side of the arrow preceded by a + sign. Using these rules, the above mechanism can be visualized in Scheme (4.99). Observing the scheme, it can be seen that the Br atom resulting in the dissociation of the upper left Br_2 molecule can react with an H_2 molecule to give an HBr molecule and an H atom. The emerging H atom can then react with a Br_2 molecule to give another HBr molecule and a Br atom. This Br atom can then react with another H_2 molecule; thus the cycle re-begins. This cycle can run over and over again, until the Br atom needed to continue the cycle would recombine, producing the lower left Br_2 molecule. If the H atom within the cycle reacts not with a Br_2 but an HBr molecule, the cycle is maintained, but this reaction consumes an HBr product, thus slowing down the process of HBr formation.[4]

---

[4]The word of Latin origin *inhibition* means stopping or slowing down a process. In this case, the second meaning is associated to the expression of inhibition reaction. It is also called *retardation*.

4.10 Chain Reactions

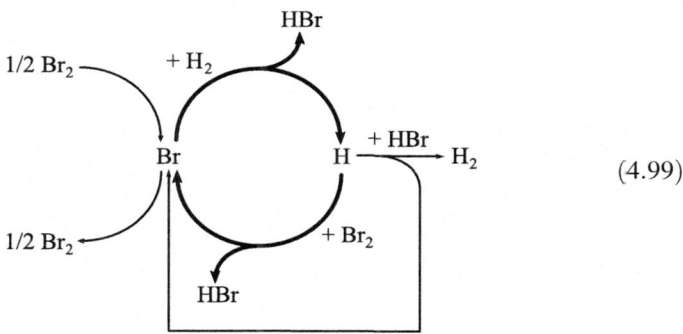

(4.99)

Scheme 4.1 Directed multigraph presentation of the chain reaction mechanism (4.94)–(4.98)

The 'chain' should be visualised in a somewhat different way: by imagining that subsequent reaction cycles get connected as the links of a chain (the expression 'chain reaction' has been coined by Max Bodenstein who has used the chain of his pocket watch to visualise this type of reaction for his students during a lecture.) The *chain length* is the number of continuously attached links before the recombination of a Br atom would terminate the chain. Atomic components H and Br are called *chain carriers* or *active centres*, as they make the chain to continue. The name *unbranched chain* (also called *straight chain*) indicates that from *one* chain carrier, there will be also *one* other chain carrier produced; thus, the cycle continuing the chain closes on its own, it is not 'branching'.

The attentive reader might have noticed that the mechanism does not contain either the dissociation of the H_2 molecule or the recombination of the H atom. The reason for this is the thermodynamics of the reactions. The dissociation energy of Br_2 is 190 kJ/mol, while that of the very stable H_2 is 430 kJ/mol. The dissociation of the former can happen (at a very slow rate) already at room temperature, but the available thermal energy is not sufficient for the dissociation of the latter (the average thermal energy is approximately the product RT; its value around room temperature is only a few kJ/mol). As the H atom can only be produced following the dissociation of Br_2, there will be much less H atom then Br in the reaction mixture; thus, its recombination is much less probable. The possible reaction Br + HBr is also missing from the mechanism, while H + HBr is there. This has also a thermodynamic rationale; the former reaction is *endothermal* with an energy demand of 170 kJ/mol, while the latter is *exothermal* releasing 67 kJ/mol. As a consequence, there are no other reaction steps contributing to the mechanism (4.94)–(4.98) than those written there – at least at moderate (not too high) temperature.

Let us see now that this mechanism is in accordance with the experimentally observed rate law (4.93). Let us first write the rate equations for every reacting component; each containing as many terms as there are reactions it is involved in.

$$\frac{d[\text{Br}_2]}{dt} = -k_1[\text{Br}_2][\text{M}] - k_3[\text{H}][\text{Br}_2] + k_5[\text{Br}]^2[\text{M}] \tag{4.100}$$

$$\frac{d[\text{H}_2]}{dt} = -k_2[\text{Br}][\text{H}_2] + k_4[\text{H}][\text{HBr}] \tag{4.101}$$

$$\frac{d[\text{Br}]}{dt} = 2k_1[\text{Br}_2][\text{M}] + k_3[\text{H}][\text{Br}_2] + k_4[\text{H}][\text{HBr}] - k_2[\text{Br}][\text{H}_2]$$
$$- 2k_5[\text{Br}]^2[\text{M}] \tag{4.102}$$

$$\frac{d[\text{H}]}{dt} = k_2[\text{Br}][\text{H}_2] - k_3[\text{H}][\text{Br}_2] - k_4[\text{H}][\text{HBr}] \tag{4.103}$$

$$\frac{d[\text{HBr}]}{dt} = k_2[\text{Br}][\text{H}_2] + k_3[\text{H}][\text{Br}_2] - k_4[\text{H}][\text{HBr}] \tag{4.104}$$

Based on Scheme (4.99) and the thermodynamic data, it is easy to see that the concentrations of diatomic molecules H_2, Br_2 and HBr are always much higher than atomic H and Br, which are produced slowly but react very quickly. For this reason, let us consider atomic species as QSS components. Following the way of expressing these conditions, let us write their rate of formation to the left-hand side and the rate of decomposition at the right-hand side by equating them.

$$2k_1[\text{Br}_2][\text{M}] + \boldsymbol{k_3[\text{H}][\text{Br}_2]} + \boldsymbol{k_4[\text{H}][\text{HBr}]} \cong \boldsymbol{k_2[\text{Br}][\text{H}_2]} + 2k_5[\text{Br}]^2[\text{M}] \tag{4.105}$$

$$\boldsymbol{k_2[\text{Br}][\text{H}_2]} \cong \boldsymbol{k_3[\text{H}][\text{Br}_2]} + \boldsymbol{k_4[\text{H}][\text{HBr}]} \tag{4.106}$$

Summing the two equations, terms in bold type cancel and we can simplify by the concentrations [M] to get the following approximation:

$$2\,k_1\,[\text{Br}_2] \cong 2\,k_5\,[\text{Br}]^2, \tag{4.107}$$

from which we can express the steady-state concentration of the Br atoms,

$$[\text{Br}] \cong \left(\frac{k_1}{k_5}[\text{Br}_2]\right)^{\frac{1}{2}}. \tag{4.108}$$

Substituting this into equation (4.106), we can solve it to get the concentration of the H atoms,

$$[\text{H}] \cong \frac{k_2[\text{H}_2]\left(\frac{k_1}{k_5}[\text{Br}_2]\right)^{\frac{1}{2}}}{k_3[\text{Br}_2] + k_4[\text{HBr}]}. \tag{4.109}$$

Let us substitute these two steady-state concentrations into equation (4.104) which gives the rate of formation of HBr, but let us also make use of the content of equation (4.103) and the fact that the rate of change of the QSS component H is approximately zero, which provides the following:

4.10 Chain Reactions

$$k_2[\text{Br}][\text{H}_2] - k_4[\text{H}][\text{HBr}] = k_3[\text{H}][\text{Br}_2]. \tag{4.110}$$

Taking this into account, the result of the substitution becomes considerably simple,

$$\frac{d[\text{HBr}]}{dt} \cong \frac{2k_2 k_3 [\text{H}_2][\text{Br}_2] \left(\frac{k_1}{k_5}[\text{Br}_2]\right)^{\frac{1}{2}}}{k_3[\text{Br}_2] + k_4[\text{HBr}]} = \frac{2k_2 \left(\frac{k_1}{k_5}[\text{Br}_2]\right)^{\frac{1}{2}}[\text{H}_2]}{1 + \frac{k_4}{k_3}\frac{[\text{HBr}]}{[\text{Br}_2]}}. \tag{4.111}$$

Let us denote the compound constant $2k_2 \left(\frac{k_1}{k_5}\right)^{1/2}$ as k and the ratio $\frac{k_4}{k_3}$ as k'; thus we get back the experimentally verified rate (4.93):

$$\frac{d[\text{HBr}]}{dt} = k \frac{[\text{H}_2][\text{Br}_2]^{1/2}}{1 + k' \frac{[\text{HBr}]}{[\text{Br}_2]}}.$$

Thus, by considering the two atomic species as QSS components, we have successfully interpreted the experimentally verified rate equations with the unbranched chain mechanism (4.94)–(4.98).

Note that by measuring the time-dependent concentration of the diatomic species H_2, Br_2 and HBr, based on the QSSA formalism, we can only determine the parameters k and k'. To determine the rate coefficients k_1–k_5 with reliable confidence, contained in the expressions $2k_2 \left(\frac{k_1}{k_5}\right)^{1/2}$ and $\frac{k_4}{k_3}$, the time-dependent concentrations of the atomic species H and Br should also be measured, and, in addition, we would also need the function providing the time dependence of their concentration which would be available by solving the system of differential equations (4.100)–(4.104). However, neither explicit nor implicit solutions exist for this system of differential equations; therefore, the parameters k_1–k_5 can only be determined by numerical integration.

Next we shall show the time-dependent concentration functions without application of QSSA, using the previously discussed numerical integration, and also with the application of QSSA. Comparing the two results, we shall see the quality of the QSSA in this case.

Diagrams in Fig. 4.7. are based on experimental results reported in the paper by Max Bodenstein and S. C. Lind with the title, 'Geschwindigkeit der Bildung des Bromwasserstoffs aus seinen Elementen', published in the issue No. 57 of *Zeitschrift für Physikalische Chemie* in 1907, pages 168–192. First we estimated the rate coefficients $k_1 - k_5$ with the help of a numerical integrator using data measured at 301.3 °C, and then we calculated the concentration profiles using these parameters[5]

---

[5]Though we cannot estimate all the five parameters $k_1 - k_5$ properly, an estimation of these parameters with most correlation coefficients close to 1 is possible. This means that three of the five rate coefficients are not unique but can change within the limits of the relations that fulfil $k = 2k_2 \left(\frac{k_1}{k_5}\right)^{1/2}$ and $k' = \frac{k_4}{k_3}$.

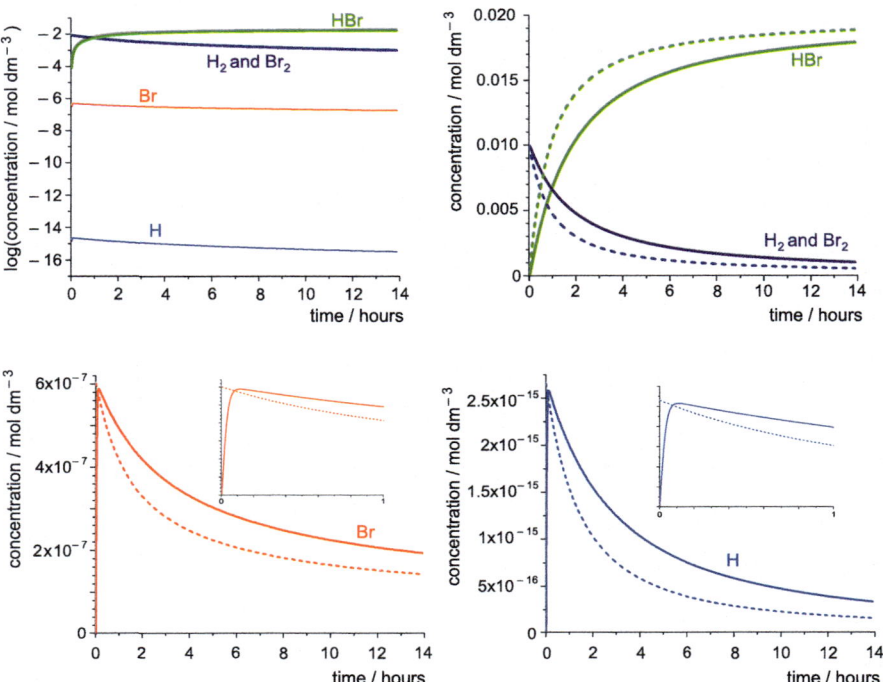

Fig. 4.7 Concentration profiles of the reaction $H_2 + Br_2 = 2\,HBr$ at 301.3 °C and high pressure (where the total concentration of the gas is 0,2 mol/dm^3). The upper left panel shows the concentrations at a ten-base logarithmic scale. The upper right diagram shows the temporal evolution of the concentration of stable diatomic species, while the two lower diagrams those of the atomic species H and Br. Insets show the concentration profiles only within the first 1 hour. It is readily seen that the concentration of chain carrier Br atoms is ten thousand times lower, than that of the diatomic species, and the concentration of the other chain carrier – H atoms – is more than ten million times lower than that. Solid lines show the results of exact numerical calculations, while dashed lines are the results of a QSS approximation as calculated from the proper rate coefficients. Note that the initial rise of Br and H atoms within the first *ca.* 10 minutes is missing in the QSS approximation

(the resulting curves perfectly fit the experimental data). Then, we calculated the two parameters $k = 2k_2 \left(\dfrac{k_1}{k_5}\right)^{1/2}$ and $k' = \dfrac{k_4}{k_3}$ of the QSSA rate law and calculated the concentration profiles according to the QSSA expressions (4.93), (4.108) and (4.109).

In the upper left diagram we can clearly see that the concentrations of Br and H atoms are lower by 4 and 12 order of magnitudes, respectively, than those of the reactants and the product. As the relatively high rate of formation of the product is only possible via chain carrier atoms, their rate of formation and decomposition should also be roughly the same as that of the product formation. The difference in the rate of formation and decomposition of the atomic species is so small that the Br atoms can only accumulate up to one millionth of one mol/dm^3, while H atoms to much less. This feature supports the use of QSSA and also explains why we should not take into account the recombination of the H atoms.

In the lower two diagrams, we can clearly see that the QSS hypothesis does not provide good quality data for the concentration of the atomic species H and Br. The most striking difference can be observed within the first 10 minutes; in this time interval, the real-life concentrations rise steeply before they begin to decrease. It is also evident that, after the first few minutes (following the rise of the concentration of atomic species), QSSA is always underestimating the concentration of atomic species.

Observing the diagrams, it is clear why we cannot formulate the QSS hypothesis in the form that the rate of concentration change for the QSS species would be zero; obviously, this is not the case. A proper formulation is that the difference in the rates of formation and decomposition is negligible compared to either of them – as expressed by equations (4.105) and (4.106).

There is another interesting feature of using the QSS approximation. If we estimate the constants k and k' by fitting the QSSA rate equation (4.93) to the experimental data, we obtain different values than what we can calculate using equations (4.108) and (4.109), using the parameters obtained by fitting the detailed mechanism (4.94)–(4.98). However, if we use the two directly estimated parameters k and k' to calculate the concentration profiles, we get practically the same curves as with the exact numerical calculations (except, of course, for the initial rise within a few minutes). This means that the QSS approximation expressed in terms of the rate equation (4.93) can be applied to get very good results for the concentration profiles, but its parameters k and k' are not identical to those we can calculate using the QSS approximation and the proper rate coefficients (this is the reason why the rate equation (4.93) suggested in the Bodenstein-Lind paper was in perfect agreement with the experimental observations).

Summing up we can state that QSSA is a very good approximation – except for the initial rise of the QSS component concentrations – if we estimate the parameters of the rate equation based on the QSSA result, not deriving them from the parameters of the more reliable mechanism. QSSA can be an important means to reduce the necessary calculation time if the reaction is too much complex. However, we should be careful if we want the best approximation with a less complicated kinetic model.

4.10.2 Branched Chain Reactions and Explosions

In case of branched chain reactions, the cyclic reaction produces more chain carriers than it consumes. As a consequence, the cycle continuing the chain not only closes on its own, but at the same time new cycles start to build, and in turn, they also produce more than one chain carriers, thus, more than one cycle starts to build again etc. As a consequence, the number of chain carriers increases exponentially, along with the rate of reaction, often leading to *explosion*.

The example shown here for a branching chain reaction is the formation of water from its elements,

$$2H_2 + O_2 = 2\,H_2O. \tag{4.112}$$

This reaction looks very simple; but in reality, its rather complex mechanism largely depends also on the pressure and temperature of the reaction mixture. To begin the discussion of the behaviour of the composite reaction, we write several elementary steps which play important role within usual conditions.

$H_2 + O_2 \rightarrow \cdot H + \cdot HO_2$	initiation	(4.113)
$\cdot OH + H_2 \rightarrow \cdot H + H_2O$	propagation	(4.114)
$\cdot H + O_2 \rightarrow \cdot OH + \cdot O\cdot$	**branching**	(4.115)
$\cdot O\cdot + H_2 \rightarrow \cdot OH + \cdot H$	**branching**	(4.116)
$\cdot H + O_2 + M \rightarrow \cdot HO_2 + M$	termination	(4.117)
$\cdot H \rightarrow$ wall	termination (1)	(4.118)
$\cdot O\cdot \rightarrow$ wall	termination (1)	(4.119)
$\cdot OH \rightarrow$ wall	termination (1)	(4.120)
$\cdot HO_2 + H_2 \rightarrow \cdot H + H_2O_2$	initiation (2)	(4.121)
$2\,\cdot HO_2 \rightarrow H_2O_2 + O_2$	termination (2)	(4.122)
$H_2O_2 \rightarrow 2\,\cdot OH$	initiation (2)	(4.123)

Reaction (4.113) is the dominant initiation step; it is thermodynamically favourable and its activation energy is not very high. Chain branching can be illustrated the simplest way by summing two propagation reaction (4.114) with the two branching steps:

$$\cdot OH + H_2 \rightarrow \cdot H + H_2O$$
$$\cdot OH + H_2 \rightarrow \cdot H + H_2O$$
$$\cdot H + O_2 \rightarrow \cdot OH + \cdot O\cdot$$
$$\underline{\cdot O\cdot + H_2 \rightarrow \cdot OH + \cdot H}$$
$$H + O_2 + 3H_2 \rightarrow 3\,\cdot H + 2H_2O$$

The sum of these steps shows that from one chain carrier H atom, three other H atoms are produced. Thus, by each of the newly initiated chains there will be again three other H atoms produced: this means that after the second run of the cycles, there will be 3^2 chain carrier H atoms. Branching will continue the same way, and as a result, the number of chain carrier species – and thus the rate of the reaction – will increase exponentially (the *branching ratio* – the number of H atoms formed from

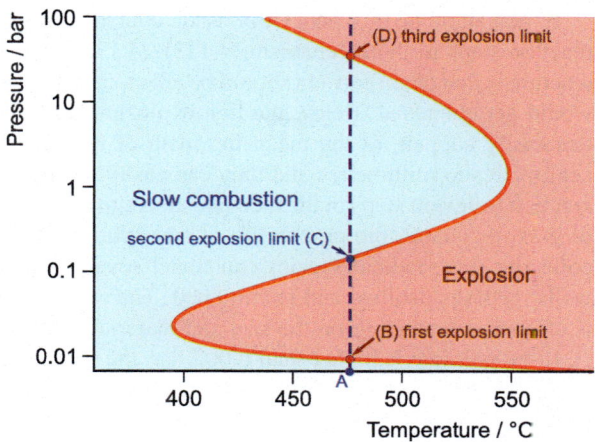

Fig. 4.8 Explosion limits diagram of the stoichiometric hydrogen–oxygen mixture as a function of pressure (logarithmic scale) and temperature (linear scale). The mixture spontaneously explodes in the pink region, while in the light-blue region, there is only combustion after ignition with a moderate propagation speed of the flame front

one H atom during one cycle – will of course be smaller than three; partly due to termination steps, partly to the fact that the rates of the four reactions summed above are different. However, the branching ratio will be higher than one anyway).

In the pressure- and temperature range where these reactions have an important contribution to the mechanism following initiation, an explosion will typically occur. The dominant termination step in this case is the formation of the hydroperoxyl radical ·HO_2 in reaction (4.117). This radical is not much reactive, thus it can easily get transported to the wall of the reactor and quenched (transformed into non-radical product). However, this termination process is not effective enough; it opens up the way to branching and eventual explosion (It is interesting to note that the termination step (4.117) competes with the branching step (4.115). However, the role of the third body – M: any stable molecule in the reaction mixture – in the former reaction suggests strong pressure dependence.) (Fig. 4.8).

Concerning explosion, the behaviour of the $H_2 + O_2$ mixture is not that simple. Depending on pressure and temperature, chains can be branched leading to explosion, or they can be unbranched when only a 'slow reaction' happens, which is usual combustion propagating in space with moderately high speed. Let us follow the pressure dependence at 480 °C, shown by the dashed line from A to D in the figure. Below the *first explosion limit* (between A and B), there is only slow combustion,[6] but at higher pressures, up to the *second explosion limit* (between B and C), the mixture explodes. Further increasing the pressure, there is again only slow combustion in a quite large pressure range (between C and D). Finally, above the *third explosion limit* (from D on), the gas mixture always explodes.

---

[6]Here, slow combustion does not mean the obsolete term which was the decomposition in physiological reactions of hydrocarbons, happening in living organisms. Here, it is the usual gas-phase flame-like combustion, which is not an explosion, rather the propagation of flame in fuel and oxygen mixtures, with the typical speed of a few tens of cm/s.

Explanation of this behaviour is of course in the pressure dependence of the reaction steps in the mechanism (4.113)–(4.123). Condition of the branched chain reaction is that chain carriers should be able to initiate enough new cycles before they would get quenched. Below the first explosion limit (B; *ca.* 0.002 bar), quenching can easily happen, as the mean free path of the radical species is in the range of centimetres to millimetres and they can easily reach the walls of the vessel without reaction (relevant steps in the mechanism are marked by (1)). In typical vessels made of glass or quartz, the probability of adsorption at the wall is lower than 0.01 per collision; thus, radical species can travel several tens of centimetres reaching the walls, getting adsorbed and recombined. The exact value of the first explosion limit is of course depending on the size and material of the vessel.

If the pressure is higher than 0.002 bar, the mean free path becomes too low and chain carriers react more often in branching steps than recombine at walls; thus, reaction steps (4.113)–(4.117) will dominate and explosion happens. This behaviour is maintained in the temperature- and pressure range where branching steps are frequent. However, above the second explosion limit marked by C (*ca.* 0.045 bar), the mean free path of the H atoms decreases below 1 µm, and they often collide with oxygen molecules, then quickly enough with a third partner so that low-reactivity ·HO_2 radicals are formed and there won't be explosion. These conditions persist in a large pressure range, up to the third explosion limit (D; *ca.* 60 bar).

If the pressure is higher than that, the mean free path of all reactive molecules falls below 100 nm. Under these conditions, collisions are very frequent and particles cannot get away from each other easily. Consequently, the released energy of the formation of an H_2O molecule cannot dissipate from the site of the reaction and the mixture is quickly heated up. At increased temperature, the rate of chain carrier formation also increases; thus the overall reaction speeds up, resulting in additional temperature increase. During this procedure, the last three steps in the mechanism marked by (2) become important, leading to explosion in this pressure range. However, the key element of this explosion is not the initiation by dissociation of H_2O_2 but local overheating; thus it is called *thermal explosion*. Conditions of thermal explosion persist at even higher pressures; thus, above the third explosion limit, explosion happens at all pressures (from this explanation it follows that there is no need for a branched chain reaction in this regime. However, it typically also accompanies thermal explosion – as it happens also in this case).

Explosion limits are also influenced by the presence of non-reactive species; *e.g.* N_2 molecules in case of hydrogen-air mixtures. These molecules can absorb excess energy from newborn product molecules with a different efficiency than reactive species, which explains their influence on the limits.

Problems

1. Consider the following two-step consecutive mechanism:

$$A_1 \xrightarrow{k_1} A_2 \xrightarrow{k_2} A_3.$$

Let us solve the rate equations of this mechanism using the eigenvalue–eigenvector method, with the following initial conditions: $c_1(t=0) = c_{1,0}, c_2(t=0) = c_3(t=0) = 0$. (*C.f.* Sect. 4.5.)

4.10 Chain Reactions

Solution: The relevant rate equations can be written as follows:

$$\dot{c}_1 = -k_1 c_1$$
$$\dot{c}_2 = k_1 c_1 - k_2 c_2$$
$$\dot{c}_3 = k_2 c_2$$

Here we have used the simplified notation for $\frac{dc_i}{dt} \equiv \dot{c}_i$, where c_1, c_2 and c_3 are components of the vector $\underline{c}$. Let us write the kinetic matrix according to Eq. (4.57):

$$\mathbf{K} = \begin{pmatrix} -k_1 & 0 & 0 \\ k_1 & -k_2 & 0 \\ 0 & k_2 & 0 \end{pmatrix}$$

The characteristic equation of matrix $\mathbf{K}$ to find the eigenvalues λ is the following:

$$\det(\mathbf{K} - \lambda \mathbf{I}) = \begin{vmatrix} -k_1 - \lambda & 0 & 0 \\ k_1 & -k_2 - \lambda & 0 \\ 0 & k_2 & -\lambda \end{vmatrix} = 0$$

Let us expand this determinant using minors and cofactors. This operation in case of a 3×3 matrix goes the following way for a general determinant in the form of

$$\begin{vmatrix} a_{11} & a_{12} & a_{13} \\ a_{21} & a_{22} & a_{23} \\ a_{31} & a_{32} & a_{33} \end{vmatrix}.$$

The *minor* $|M_{ij}|$ for an element a_{ij} of the determinant of a matrix $\mathbf{A}$ is the determinant that results when the row and column that element a_{ij} is in are deleted. The *value* of the determinant can be calculated adding terms of products of the following structure:

$$\det \mathbf{A} = \sum_{j=1}^{3} (-1)^{i+j} a_{ij} |M_{ij}|.$$

Thus, the value of the determinant of a 2×2 matrix is the product of the elements on the main diagonal minus the product of the elements off the main diagonal. For example, the value of the minor $|M_{11}| = \begin{vmatrix} a_{22} & a_{23} \\ a_{32} & a_{33} \end{vmatrix}$ is $a_{22}a_{33} - a_{23}a_{32}$.

Applying these rules, we get $\det(\mathbf{K} - \lambda \mathbf{I}) = (-k_1 - \lambda)(-k_2 - \lambda)(-\lambda) = 0$, which is the *characteristic equation* of the kinetic matrix $\mathbf{K}$. The eigenvalues of $\mathbf{K}$ are solutions of this equation:

$$\lambda_1 = -k_1; \quad \lambda_2 = -k_2; \quad \lambda_3 = 0.$$

Now we have to find the eigenvectors $\underline{r}$ for each eigenvalue. Substituting eigenvalue $\lambda_1 = -k_1$ into the original system of equations we get:

$$0 \cdot r_1 + 0 \cdot r_2 + 0 \cdot r_3 = 0$$
$$k_1 \cdot r_1 + (-k_2 + k_1) \cdot r_2 + 0 \cdot r_3 = 0$$
$$0 \cdot r_1 + k_2 \cdot r_2 + k_1 \cdot r_3 = 0$$

The matrix of the above system of equations is

$$\boldsymbol{L} = \begin{pmatrix} 0 & 0 & 0 \\ k_1 & -k_2 + k_1 & 0 \\ 0 & k_2 & k_1 \end{pmatrix}$$

This is a *degenerate* matrix, as its first raw contains only zeros. In this case, there is no unique solution of the system of equations; only the ratio of the components r_1: r_2: r_3 can be obtained as a solution. The ratios can be given as the ratios of the minors for the elements of a row – with similar alternating signs as in the case of the expansion of the determinant:

$$r_1 : r_2 : r_3 = (-1)^{i+1} \cdot |L_{i1}| : (-1)^{i+2} \cdot |L_{i2}| : (-1)^{i+3} \cdot |L_{i3}|.$$

Let us write the ratios according to the first row of $\boldsymbol{L}$ (*i.e.* setting $i = 1$):

$$r_1 : r_2 : r_3 = + \begin{vmatrix} (-k_2 + k_1) & 0 \\ k_2 & k_1 \end{vmatrix} : - \begin{vmatrix} k_1 & 0 \\ 0 & k_1 \end{vmatrix} : + \begin{vmatrix} k_1 & (-k_2 + k_1) \\ 0 & k_2 \end{vmatrix} =$$

$$= (-k_2 + k_1)k : -k_1^2 : k_1 k_2 = 1 : \frac{k_1}{k_2 - k_1} : -\frac{k_2}{k_2 - k_1}.$$

(As the eigenvector can be multiplied by any constant and it still remains an eigenvector, we can give the simplest result by setting one component to be unit.)

The eigenvector $\underline{x}$ can be obtained with $\lambda_1 = -k_1$ in the exponent:

$$x_1 = e^{-k_1 t}; x_2 = \frac{k_1}{k_2 - k_1} e^{-k_1 t}; x_3 = -\frac{k_2}{k_2 - k_1} e^{-k_1 t}.$$

Thus, the eigenvector corresponding to the eigenvalue $\lambda_1 = -k_1$ is

$$\underline{x}_1 = \begin{bmatrix} 1 \\ \dfrac{k_1}{k_2 - k_1} \\ -\dfrac{k_2}{k_2 - k_1} \end{bmatrix} \cdot e^{-k_1 t}$$

4.10 Chain Reactions

We can calculate the eigenvector corresponding to the eigenvalue $\lambda_2 = -k_2$ in a similar way. The matrix of the relevant system of equations to solve is the following:

$$L = \begin{pmatrix} -k_1 + k_2 & 0 & 0 \\ k_1 & 0 & 0 \\ 0 & k_2 & k_2 \end{pmatrix}$$

Let us write the ratios of the eigenvector components this time also according to the first row of L:

$$r_1 : r_2 : r_3 = + \begin{vmatrix} 0 & 0 \\ k_2 & k_2 \end{vmatrix} : - \begin{vmatrix} k_1 & 0 \\ 0 & k_2 \end{vmatrix} : + \begin{vmatrix} k_1 & 0 \\ 0 & k_2 \end{vmatrix} = 0 : -k_1 k_2 : k_1 k_2 = 0 : -1 : 1.$$

Thus, the eigenvector corresponding to the eigenvalue $\lambda_2 = -k_2$ is

$$\underline{x}_2 = \begin{bmatrix} 0 \\ -1 \\ 1 \end{bmatrix} \cdot e^{-k_2 t}.$$

The matrix of the system of equations corresponding to the eigenvalue $\lambda_3 = 0$ is the following:

$$L = \begin{pmatrix} -k_1 & 0 & 0 \\ k_1 & -k_2 & 0 \\ 0 & k_2 & 0 \end{pmatrix}$$

The ratios of the eigenvector components are

$$r_1 : r_2 : r_3 = + \begin{vmatrix} -k_2 & 0 \\ k_2 & 0 \end{vmatrix} : - \begin{vmatrix} k_1 & 0 \\ 0 & 0 \end{vmatrix} : + \begin{vmatrix} k_1 & -k_2 \\ 0 & k_2 \end{vmatrix} = 0 : 0 : k_1 k_2 = 0 : 0 : 1.$$

Thus, the eigenvector corresponding to the eigenvalue $\lambda_3 = 0$ is

$$\underline{x}_3 = \begin{bmatrix} 0 \\ 0 \\ 1 \end{bmatrix} \cdot e^{0t} = \begin{bmatrix} 0 \\ 0 \\ 1 \end{bmatrix}.$$

The general solution of the rate equations is the linear combination of the above three eigenvectors. The particular solution in agreement with the initial conditions can be obtained by substituting $t = 0$ into the vector $\underline{x}_i$, and setting the result equal to $\underline{c}(t = 0)$:

$$\begin{bmatrix} c_{1,0} \\ 0 \\ 0 \end{bmatrix} = b_1 \begin{bmatrix} 1 \\ \dfrac{k_1}{k_2-k_1} \\ -\dfrac{k_2}{k_2-k_1} \end{bmatrix} \cdot e^0 + b_2 \begin{bmatrix} 0 \\ -1 \\ 1 \end{bmatrix} \cdot e^0 + b_3 \begin{bmatrix} 0 \\ 0 \\ 1 \end{bmatrix} \cdot e^0$$

Let us write this as a system of three equations:

$$c_{1,0} = b_1 \quad + \quad 0 \quad +0$$

$$0 = b_1 \frac{k_1}{k_2-k_1} - b_2 + 0$$

$$0 = b_1 \frac{-k_2}{k_2-k_1} + b_2 + b_3$$

From the first equation we get $b_1 = c_{1,0}$. Substituting this into the second equation, we get $b_2 = c_{1,0} \frac{k_1}{k_2-k_1}$. Substituting both into the third equation, we get $b_3 = c_{1,0}$. Having obtained the linear combination coefficients b_1, b_2 and b_3 conform to the initial conditions, we can write the particular solution in the form of the following vector:

$$\underline{c} = c_{1,0} \begin{bmatrix} 1 \\ \dfrac{k_1}{k_2-k_1} \\ -\dfrac{k_2}{k_2-k_1} \end{bmatrix} \cdot e^{-k_1 t} + c_{1,0} \frac{k_1}{k_2-k_1} \begin{bmatrix} 0 \\ -1 \\ 1 \end{bmatrix} \cdot e^{-k_2 t} + c_{1,0} \begin{bmatrix} 0 \\ 0 \\ 1 \end{bmatrix}.$$

Alternatively, we can write individual components of this vector – the concentrations of the species A_1, A_2 and A_3 taking part in the reaction:

$$c_1 = c_{1,0} \cdot e^{-k_1 t}$$

$$c_2 = c_{1,0} \frac{k_1}{k_2-k_1} \left(e^{-k_1 t} - e^{-k_2 t} \right)$$

$$c_3 = c_{1,0} \left(1 - \frac{k_2}{k_2-k_1} \cdot e^{-k_1 t} + \frac{k_1}{k_2-k_1} \cdot e^{-k_2 t} \right)$$

This solution method is perhaps more complicated than usual integrations shown in Sect. 4.3; but in case of more than two consecutive steps, it becomes substantially simpler. Its great advantage is that the procedure itself is not more complicated for any number of reaction steps than the one shown here. Recently, eigenvalue–eigenvector problems can easily be solved using symbolic computation programs (*e.g.* Mathematica, Maple, MathCad). However, to our experience, these programs cannot easily deal with large matrices containing many zero elements – though most of the kinetic matrices do have this property. In such cases, the programs need some

4.10 Chain Reactions

'expansion' of the problem to solve. Thus, it is recommended to use them only if an explicit solution is needed.

2. In a recent study (see Further Reading; Horváth et al.), an Exenatide-analogue peptide was reduced, in which two cysteines are coupled by an intermolecular S–S bond. (Exenatide is a diabetes drug.) Reduction by TCEP (tris(2-carboxyethyl)-phosphine; a reducing agent frequently used in biochemistry) resulted in breaking the S–S bond and reconstructing non-bonded S–H groups in the two cysteines. However, the solvent (water) contained some dissolved oxygen, which re-oxidized part of the reduced polypeptide until the oxygen was consumed completely. The kinetic model mechanism was found to be the following:

$$SS + RA_{red} \xrightarrow{k_1} 2SH + RA_{ox}$$
$$2SH + O_2 \xrightarrow{k_2} SS,$$

where SS is the peptide with the S–S bond-coupled cysteins, RA_{red} is the reducing agent TCEP, 2SH is the reduced peptide with the two SH groups on the cysteines, and RA_{ox} is the oxidized form of the reducing agent.

In a kinetic experiment, the following initial conditions were realised:
The initial concentration of RA_{red} was 3.257 mmol/dm^3, that of RA_{ox} and 2SH was zero.

The following concentrations of the species SS and 2SH as a function of reaction time were determined by NMR measurements of selective proton signals (see Sect. 7.1):

time (min)	0	150	400	650	900	1150	1400	1650	1900	2150	2400
[SS] (mmol/dm^3)	1.486	1.388	1.079	0.859	0.770	0.665	0.688	0.585	0.595	0.521	0.487
[2SH] (mmol/dm^3)	0.0	0.081	0.324	0.467	0.558	0.718	0.820	0.859	1.078	1.109	1.137

time (min)	2650	2900	3150	3400	3650	3900	4150	4400	4650	4905
[SS] (mmol/dm^3)	0.407	0.380	0.348	0.308	0.275	0.312	0.277	0.221	0.222	0.199
[2SH] (mmol/ dm^3)	1.205	1.234	1.238	1.158	1.160	1.346	1.353	1.320	1.394	1.429

Using the above data and a kinetic integrator software, determine the rate coefficients k_1 and k_2, as well as the unknown initial concentration of O_2, and that of the species SS – which was not controlled at a good precision.

Solution: Let us choose a reaction kinetic integrator software first. At the time of writing this book, a freely available and versatile software is COPASI (http://copasi.org/; a biochemical system simulator) which we have used to solve this problem. If the model contains mechanistic steps with mass action kinetic law (like in the present problem), the kinetic model can be entered into the user interface in the form of stoichiometric equations. In addition to the reaction steps, we also have to enter the volume of the reactor, the species and their initial concentrations, and relevant kinetic parameters, along with their units. We also have to give the path and the name of the data file, along with a specification of its content.

To determine suitable initial values of the parameters to estimate, we can use the simulation option of the software called 'Time course'. Starting with some 'intelligent guess' of the parameters, we can simulate the time course of the reaction and observe the resulting temporal evolution of the concentrations of the measured species. For this option, we should also specify the duration in time of the simulation and the time intervals to use for the numerical integration, along with the integration method. We can choose among several integration methods (explained in the user manual), and set tolerance criteria to stop iterative integration steps. Comparing the results with the shape of the experimental temporal evolution, we can accordingly change parameters and repeat the simulation until the shape of the simulated curves have a fair similarity to the shape in the experimental results (it typically takes only a few runs). Then, we should use the option 'Parameter estimation'. Here we can assign data sets to concentrations of species, and also choose which parameters should be estimated (other parameters should be fixed). We should then set the initial values of the parameters to be estimated. We can also choose among several parameter estimation methods (explained in the user manual), and set tolerance criteria for the convergence of the iterative determination of the parameters. We can also choose output options. A rather useful option is to plot the measured data along with the fit of the model using the estimated parameters, and, of course, to provide a detailed statistical inference concerning the estimation as well. When doing all this, we can run the parameter estimation and evaluate the graphical and numerical results obtained. In case we are content with the estimation and fitting, the procedure is finished and we can save the output. If this is not the case, we could change several options and initial parameters and re-run the parameter estimation procedure.

In the case of the present problem – which is relatively simple to solve – we have used the integration method LSODA (Livermore Solver for Ordinary Differential equations with Automatic switching between stiff and non-stiff solver) both for simulations and parameter estimation, with a tolerance of 1×10^{-12} (a difference between subsequent integrated values to stop the iteration). Preliminary time course simulations have shown that the shapes of the simulated curves fairly mimic that of the experimental ones with the parameter values $k_1 = 0.008$ dm^3 mol^{-1} min^{-1}; $k_2 = 0.0001$ dm^3 mol^{-1} min^{-1}; $[SS]_o = 1.5$ mmol/dm^3 (this latter close to the first measured value); $[O_2]_o = 0.5$ mmol/dm^3 (close to the literature value of the equilibrium concentration of oxygen in pure water). We have chosen these values as starting parameters for the Levenberg–Marquardt estimation method, and a tolerance of 1×10^{-6} (a difference between the sum of squared residual errors in subsequent iteration steps to stop the procedure). Using these settings, the first run of the estimation gave the following (reliable and satisfactory) results.

4.10 Chain Reactions

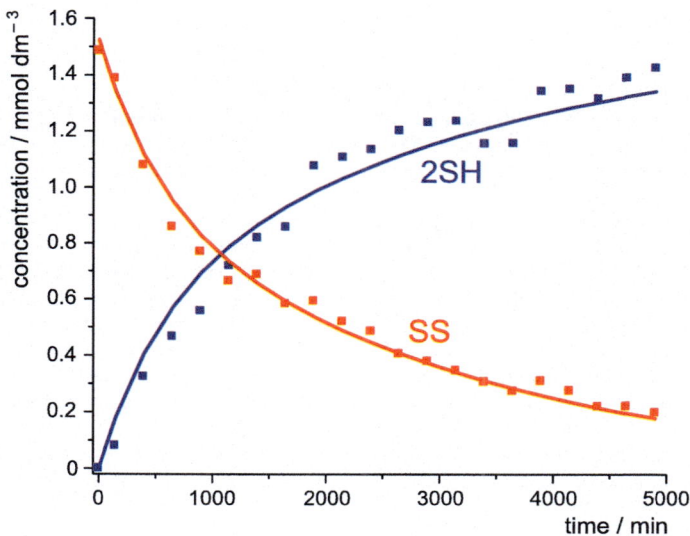

Fig. 4.9 Quality of the fit of the model mechanism with the estimated parameters. Squares show measured data, continuous curves the fitted concentrations with the estimated parameters

Parameter	Estimated value	Standard deviation	95% confidence interval half width
k_1 ($dm^3\ mol^{-1}\ min^{-1}$)	2.71×10^{-4}	1.96×10^{-5}	4.82×10^{-5}
k_2 ($dm^3\ mmol^{-1}\ min^{-1}$)	7.26×10^{-4}	1.93×10^{-4}	4.74×10^{-4}
$[SS]_o$ ($mmol/dm^3$)	1.527	0.013	0.031
$[O_2]_o$ ($mmol/dm^3$)	0.336	0.073	0.181

Confidence intervals shown have been calculated using Student's t-distribution of the estimated parameters with 17 degrees of freedom (21 data minus four parameters), the actual critical value at 95% confidence being 2.458.

Figure 4.9 shows the quality of the fit of the model mechanism with the estimated parameters (it is the same diagram which has been drawn by COPASI). Squares show measured data, while continuous curves the fitted concentrations with the estimated parameters.

Summing up we can conclude that the model mechanism fits the measured data well and that the parameters k_1 and $[SS]_o$ – reduction rate coefficient and the initial concentration of the S–S bonded peptide – are well determined; while the other two parameters – back-oxidation rate coefficient k_2 and the initial concentration of dissolved oxygen $[O_2]_o$ – are also determined, but only within substantially larger confidence intervals.

Further Reading

1. Pilling MJ, Seakins PW (1995) Reaction kinetics. Oxford University Press, Oxford
2. de Paula J, Atkins PW (2014) Physical chemistry. 10th edn. Oxford University Press, Oxford
3. Silbey LJ, Alberty RA, Moungi GB (2004) Physical chemistry. 4th edn. Wiley, New York
4. Steinfeld JI, Francisco JS, Hase WL (1998) Chemical kinetics and dynamics. 2nd edn. Prentice Hall, Englewood Cliffs
5. Press WH, Teukolsky SA, Vetterling WT, Flannery BP (2007) Numerical recipes: the art of scientific computing. 3rd edn. Cambridge University Press, Cambridge
6. Green JR, Margerison D (1978) Statistical treatment of experimental data. Elsevier, Amsterdam
7. Bevington PR, Robinson DK (2002) Data reduction and error analysis for the physical sciences. 3rd edn. McGraw-Hill, New York
8. Horváth D, Taricska N, Keszei E, Stráner P, Farkas V, Tóth GK, Perczel A (2019) Compactness of protein folds alters disulfide-bond reducibility by three orders of magnitude: a comprehensive kinetic case study on the reduction of differently sized tryptophan cage model proteins. ChemBioChem, Wiley-VCH 21:681–695
9. COPASI: Biochemical system simulator. http://copasi.org/; last access 16 Oct 2020

Chapter 5
Activation Processes and Unimolecular Gas Phase Reactions

Molecular theories of chemical reactions have been described in Chap. 2 where both collision theory and transition state theory (TST) were developed for gas phase reactions comprising two molecules. However – as it is also mentioned there – we can generalise TST for *unimolecular* reactions as well. Most prominent types are *isomerisation* and *spontaneous dissociation* reactions. After the general treatment of bimolecular reactions, we can ask for the explanation of reactions where bond breaking or internal rearrangement of single molecules happens. It is obvious that molecules undergoing these reactions should have enough energy being able to rearrange, or to put it simply, the activation energy should be 'within the molecules' reacting. It is also easy to recognise that the activation energy should reside in the critical mode of the molecule which leads to the actual reaction.

In this chapter, we shall first discuss the ways molecules can gather the activation energy; then we analyse in detail the case when the source of activation energy is a collision with another molecule.

5.1 Molecular Interpretation of Activation

Activation is the process when a reacting molecule (or molecules in case of more than one partner) acquires enough energy from some kind of energy source which makes a reaction possible. The source of energy may be quite diverse in its nature. In Chap. 2 we assumed that the activation energy is involved in the kinetic energy of the two colliding molecules. In case of a unimolecular reaction, it is trivial that only those molecules can react whose energy exceeds the activation energy barrier. However, these high-energy molecules get transformed and disappear following their reaction; thus, there should be a continuous supply of the excess energy in order to maintain the reaction. This procedure of 'charging' molecules with enough energy can happen by different means; accordingly, we distinguish several methods of activation.

Recently, in more and more kinetic studies activation is done using light (*i.e.* photons), which is called *photoactivation* or *photochemical activation*. The key element of this is the absorption of one or more photons by molecules to overcome the activation barrier. There exists a distinct discipline called *photochemistry* which deals with photoactivation and subsequent chemical reactions. However, photochemistry comprises only activations with relatively low energy photons, limited to processes where the absorption of one or a few photons provides just enough energy for a molecule to react. The feasibility and flexibility of many types of reactions to initiate by the absorption of photons were made possible by the discovery and quick development of lasers. Experimental methods in the field of laser chemistry are discussed in some detail in Chap. 7.[1]

Prior to the availability of lasers, radioactive sources played an important role in the activation of reactions. Similar to photochemistry, another discipline called *radiation chemistry* deals with activation using radioactivity as the source of energy. Besides alpha-, beta- and gamma-rays, irradiation by accelerated positrons or electrons (the latter equivalent to beta-rays), and in general, the use of any accelerated particles in the 100 eV–10 GeV energy range is also considered as radiation chemistry. Radiation used for activation considered within this discipline has the characteristic property that activating particles have energies that largely exceed reaction enthalpies and activation energies. Consequently, they initiate the reaction of a great number of molecules within a small spatial region, while continuously losing their energy. Most of this energy is typically used for ionisation of molecules; for this reason, the radiation emitted by the above-mentioned sources is also called *ionising radiation*. Note that gamma rays contain also photons but of such enormously high energies that reactions initiated by these photons do not happen by absorption of one photon per molecule, but by a series of energy losses resulting in many ionisation and other reaction events. By accelerating charged particles (*e.g.* electrons and positrons) in electromagnetic fields, short pulses can also be produced; this was the reason for the development of radiation chemistry prior to the discovery of pulsed lasers, especially from the early 1960s to the late 1990s. Many radical-molecule and ion-molecule reactions in the time range of 10^{-3}–10^{-8} s have been first studied using methods of the so-called *pulse radiolysis*.

Reactions activated by *microwave radiation* are not initiated directly by photons of few hundred megahertz to few hundred gigahertz frequency; their energy is in the order of 0.01 kJ/mol only, too small to activate reactions. The effect of microwaves is a consequence of absorption of a great number of photons highly exciting molecular rotation (or, in condensed phase, the limited rotation called *libration*). Rotational excitation then propagates into practically all molecular modes resulting in a simple heating effect. Microwave heating is very efficient as molecules in the

---

[1]The first laser was built by the American engineer-physicist Theodor Harold Maiman (1927–2007) in 1960. Laser techniques have then developed quickly, and pulsed lasers have enabled very efficient reaction kinetic applications. Most lasers used in kinetic studies emit very short, nearly monochromatic, and coherent light pulses in the UV-visible and near infrared frequency range of duration of 10^{-6}–10^{-14} s.

bulk of the reaction mixture are heated simultaneously with surface molecules, unlike the effect of an external heat source. However, this kind of activation is nothing more but thermal activation.

Many reactions can be activated by mechanical effects (*e.g.* shaking or hitting). Such reactions belong to the discipline of *mechanochemistry* or *sonochemistry*. The latter name refers to ultrasound; its loudness can easily be controlled and it is capable to activate many reactions. Ultrasonic treatment of condensed phase mixtures is called *sonication*. The wavelength of commonly used ultrasound generators of a few MHz frequency is in the millimetre range; therefore, their absorption does not occur at molecular but macroscopic size. This is the reason that sonochemical excitation is efficient in condensed phase, rather than in a gas. In liquids, ultrasound waves produce very small cavities due to interference phenomena (which is called *cavitation*). In these small cavities, a lot of energy is accumulated; as a consequence, when these small bubbles collapse, there will be important temperature and pressure increase in very small volume elements which in turn initiates the reaction. The difference with respect to microwave activation is that energy deposition is not homogeneous within the bulk of the liquid but it is confined to very small nodules only.

Another form of activation is called *chemical activation*. It is the formation of highly vibrationally excited products of (usually bimolecular association) reactions which can further react as the activation energy for the second reaction is stored in the vibrational excitations.[2]

In the remaining part of this chapter, we shall only discuss thermal excitation in gas phase reactions.

5.2 Theories of Unimolecular Gas Reactions

Unimolecular gas reactions have remarkable pressure dependence: they are second order at low pressure and first order at high pressure. Satisfying interpretation of these reactions has only been developed slowly, as a result of accumulating experimental evidence and developing theoretical background. For this reason, we also follow the traditional way of discussing subsequent theories in a historical order, associating to theories the names of kineticists who have developed them.

---

[2]The process of treating active carbon surfaces is also called chemical activation. The one considered here is a different phenomenon.

5.2.1 Lindemann Mechanism

The first successful attempt to explain the behaviour of unimolecular reactions has been conceived by Lindemann[3] who had some interest in high atmosphere chemistry. He proposed that activation (and also de-activation) of molecules happens via collisions, and molecules having acquired more than the threshold energy for activation can undergo reaction. The rate coefficient for the overall reaction emerges from these two procedures. From a kinetic point of view, we can formulate this as a composite reaction with the following mechanism:

$$A + A \underset{k_{-1}}{\overset{k_1}{\rightleftharpoons}} A^* + A \tag{5.1}$$

$$A^* \overset{k_2}{\longrightarrow} \text{products} \tag{5.2}$$

For the sake of simplicity, the mechanism is written here only in terms of the reactant A, which is only valid at the very beginning of the reaction; with the advancement of the reaction, there will be also product molecules present which can also deliver or absorb energy in reactions similar to (5.1). In a generally valid mechanism, the usual symbol [M] should be written in place of [A] as a collision partner, standing for any other molecule. Thus, the following formulae are strictly valid only for the initial rate of the reaction if the initial mixture contains only the pure substance A. However, this feature does not confound the explanation of the nature of the reaction but simplifies actual formulae.

Lindemann assumed that molecules having energy greater than the activation barrier E_0 (let us call them *energetic molecules*, denoted by A*) are quasi-stationary state (QSS) components. Accordingly, let us equate their rate of formation to the sum of the rates of reverse reaction (5.1) – losing their excess energy in collisions – and reaction (5.2) – their transformation into products – then express the QSS concentration from this equation:

$$k_1[A]^2 \cong k_{-1}[A^*][A] + k_2[A^*], \tag{5.3}$$

from which:

$$[A^*] \cong \frac{k_1[A]^2}{k_{-1}[A] + k_2}. \tag{5.4}$$

Let us substitute this into the rate equation of the product formation:

---

[3]Frederick Alexander Lindemann (1886–1957) was an English physicist of German origin. As scientific advisor of Winston Churchill during World War II, he also had an influence on British policy. He proposed the first theory of unimolecular reactions in 1922 that has later been named after him.

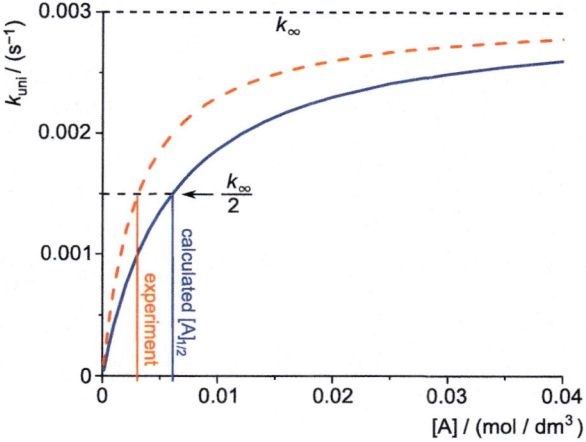

Fig. 5.1 Pressure dependence of the unimolecular rate coefficient based on experimental data and calculated from the Lindemann mechanism. (The pressure of a gas containing only component A is proportional to the concentration [A].) Continuous blue curve shows k_{uni} calculated from Eq. (5.9), while the dashed red curve is fitted to experimental data. Corresponding values of $[A]_{1/2}$ are also shown with the same colours

$$\frac{d[\text{products}]}{dt} \cong k_2[A^*] = \frac{k_1 k_2 [A]^2}{k_{-1}[A] + k_2}. \tag{5.5}$$

Compare this result with the formal first-order rate equation of the reaction $A \xrightarrow{k_{\text{uni}}} $ products:

$$\frac{d[\text{products}]}{dt} = k_{\text{uni}}[A]. \tag{5.6}$$

As a result of the comparison, we can express the rate coefficient k_{uni} using parameters originating from Eq. (5.3), but the QSS concentration of A* does not appear in the expression:

$$k_{\text{uni}} = \frac{k_1 k_2 [A]}{k_{-1}[A] + k_2}. \tag{5.7}$$

Figure 5.1. shows the pressure dependence of the above expression. (Pressure in this case is proportional to the concentration [A].) We can see that the shape of the calculated curve is roughly the same as the one showing pressure dependence of the experimentally observed k_{uni}; it is a linear function of [A] at low pressure (thus the reaction is of second order), while at high enough pressure (typically above 1 bar), it becomes slowly saturated (independent of the concentration [A]; thus the reaction is of first order). The pressure range between these two limiting regions is usually called the *fall-off region*.

Limiting behaviour at low and high pressures can also be explained by considering limiting values of Eq. (5.7). At low enough pressures where $k_{-1}[A] \ll k_2$ holds, the first term in the denominator becomes negligible compared to k_2, which provides the low-pressure limit $k_0 = k_1[A]$. Substituting this into rate Eq. (5.6) we can see that it provides a second-order rate equation. Based on the condition $k_{-1}[A] \ll k_2$ we can also conclude that the *rate determining step is the activation of molecules* in the process of acquiring energy during collisions.

At high enough pressures where $k_{-1}[A] \gg k_2$ holds, k_2 can be neglected compared to the product $k_{-1}[A]$, which – after simplifications – provides the high-pressure limit $k_\infty = \frac{k_1 k_2}{k_{-1}}$. Thus, the rate coefficient does not depend on the concentration which indicates a first-order reaction. Based on the condition $k_{-1}[A] \gg k_2$ we can conclude that, under these conditions, the *rate determining step is the reaction of active molecules* in the unimolecular process.

5.2.2 Lindemann-Hinshelwood Mechanism

As we can see, Lindemann's explanation reflects very well the overall behaviour of unimolecular gas reactions in the entire pressure range. However, when it was compared to experimental data of several reactions, it turned out that actual formulae derived from the Lindemann mechanism provide rather inaccurate results for the rate coefficient. To understand this inconsistency better, let us make the following considerations. First we divide both numerator and denominator of expression (5.7) by the product $k_{-1}[A]$:

$$k_{\text{uni}} = \frac{\frac{k_1 k_2}{k_{-1}}}{1 + \frac{k_2}{k_{-1}[A]}}, \qquad (5.8)$$

then we replace $\frac{k_1 k_2}{k_{-1}}$ with the previously defined limit k_∞:

$$k_{\text{uni}} = \frac{k_\infty}{1 + \frac{k_\infty}{k_1[A]}}. \qquad (5.9)$$

We can see that the rate coefficient k_{uni} can be calculated from the experimental k_∞, k_1 (what can be calculated using Eq. (2.7) based on collision theory[4]), and the actual concentration [A]. It is easy to see that the limiting behaviour of this formula is correct; it provides k_∞ at infinitely high and $k_0 = k_1[A]$ at infinitely low pressure.

---

[4]We expect that collision theory gives quite a good approximation for the collisions having an energy higher than E_a, as in this case we only calculate the rate of formation of energetic molecules, not that of a bimolecular reaction. However, later we shall see that the rate of activation also depends on the structure of the molecule, which is neglected in the collision theory treatment.

5.2 Theories of Unimolecular Gas Reactions

However, calculated values in the fall-off region are often different from experimental k_{uni} values by more than an order of magnitude.

This flaw of the theory can also be stated by testing a single experimental value. The concentration $[A]_{1/2}$ at which the value of k_{uni} becomes $\frac{k_\infty}{2}$ can be determined from a few experiments, and it can also be calculated using Eq. (5.9), which is based on Lindemann theory. Calculated $[A]_{1/2}$ value is obtained when $\frac{k_\infty}{k_1[A]} = 1$, thus $[A]_{1/2} = \frac{k_\infty}{k_1}$. Experimentally determined $[A]_{1/2}$ values are considerably lower than calculated ones; accordingly, experimentally determined k_{uni} values in the fall-off region are much higher than the calculated ones. As it can be seen in Fig. 5.1, this behaviour persists in the entire fall-off region. Experimental evidence also shows that the extent of the mismatch between the two values depends on the structure of the molecules A.

As k_∞ is an experimental value (truly reflecting the nature of the reaction), the reason behind the mismatch is due to an overestimation of the rate coefficient k_1. Calculation of this quantity is based on collision theory, which yields the expression for the rate coefficient of the formation of energetic molecules according to reaction (5.1) as:

$$k_1 = d^2 \pi \sqrt{\frac{8 k_B T}{\pi \mu}} e^{-\frac{E_a}{RT}}. \tag{5.10}$$

Let us denote the constant pre-exponential term – the *collision factor* – by Z_1, which contains only temperature, masses of molecules considered as elastic hard-spheres (in the form of their reduced mass μ) and their geometric size (diameter d). The expression for k_1 is thus simplified to:

$$k_1 = Z_1 e^{-\frac{E_0}{RT}}, \tag{5.11}$$

containing the energy E_0 that should be included in the molecules ready to react. Due to the conditions of validity of this expression, the energy of molecules originates in the relative kinetic energy of the colliding partners that is calculated from their relative velocity, and this should be higher than E_0.

Hinshelwood[5] upgraded the theory by stating that there is no need during the collisions to acquire so much kinetic energy; low-energy molecules A undergoing reaction (5.1) do not have zero internal energy prior to energising collisions, but higher. Therefore, their internal energy should be added to the total energy in the energetic molecule after collision. Internal energy of molecules prior to energising collision depends on the number of internal molecular modes. As unimolecular

---

[5]Cyril Norman Hinshelwood (1897–1967) was an English physical chemist. He made further development to the Lindemann mechanism in 1926. His other most important contributions to reaction kinetics were the development of the theory of heterogeneous reactions and the quantitative treatment of chain reactions. He has got the 1956 Nobel Prize in Chemistry shared with Nikolay Nikolayevich Semenov "for their researches into the mechanism of chemical reactions".

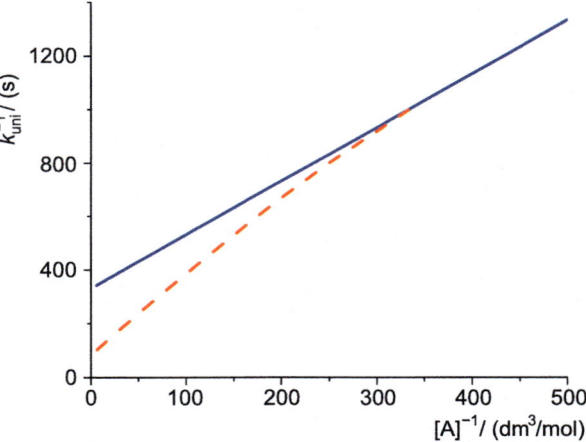

Fig. 5.2 Pressure dependence of the unimolecular rate coefficient calculated from the Lindemann mechanism. The figure shows the high-pressure behaviour. The continuous blue line shows $1/k_{uni}$ calculated from Eq. (5.13) as a function of $1/[A]$. The dashed red curve is fitted to experimental data

reactions – like isomerisation or dissociation – are connected to molecular vibrations, the key quantity is the number of vibrational degrees of freedom. To describe the situation, Hinshelwood made classical statistical mechanical calculations assuming vibrational degrees of freedom associated to vibrations of identical frequencies. His results for a molecule having s vibrational degrees of freedom were the following:

$$k_1 = \frac{Z_1}{(s-1)!} \left(\frac{E_0}{RT}\right)^{s-1} e^{-\frac{E_0}{RT}} \qquad (5.12)$$

This expression and Eq. (5.9) for k_{uni} together give the rate coefficient of the composite reaction according to the Lindemann-Hinshelwood mechanism. Though it includes the dependence of the rate coefficient k_1 on the initial energy of molecule A prior to the energising collision, it does not take into account a similar dependence of the unimolecular rate coefficient k_2 – though we expect this dependence on the basis of the same arguments as what has been used in case of k_1. This inconsistency can be shown in the following way. Let us write the inverse of the unimolecular rate coefficient from Eq. (5.9):

$$\frac{1}{k_{uni}} = \frac{1}{k_\infty} + \frac{1}{k_1} \frac{1}{[A]}. \qquad (5.13)$$

According to this expression, the inverse of k_{uni} as a function of $1/[A]$ should be linear with an intercept of $\frac{1}{k_\infty}$ and slope of $1/k_1$. However – as we can see in Fig. 5.2 – at high pressures (low $1/[A]$ values) this function considerably deviates from linearity downwards, which means lower and lower values of k_{uni} with increasing pressure. As this behaviour occurs at the high pressure limit, it is obvious that some correction similar to the above explained one concerning k_1 should also be applied in case of k_2. The theory of unimolecular reactions has been further developed into this direction by three kineticists: Rice, Ramsperger and Kassel.

5.2.3 RRK Theory

To overcome the inconsistency explained above, Rice and Ramsperger proposed classical statistical physics calculations, while Kassel[6] derived corrections based on the energy distribution in the reactant molecule using quantum physics and combinatorial calculations. As both the underlying principles and the results of the two different approaches are equivalent, this theory is named after the initials of the three authors as *RRK theory*.

The key idea of this theory is that the unimolecular transformation of the reactant molecule is not a simple elementary step. Once the energetic molecule A* has been formed via collision, an internal vibrational-energy redistribution (IVR) should happen, resulting in the transition state of the molecule, which in turn can decay into products within a single period of the critical vibration. The time for this IVR to happen depends on the actual internal energy and the structure of the molecule: the greater the energy of the molecule, the sooner it can rearrange; however, the more complicated its structure, the longer it takes for the vibrational energy to localise in the transition state configuration. (To put it simply: larger molecules having more complicated structure can survive longer prior to their decay even in case of sufficient activation energy, than smaller molecules having simpler structure.) The composite mechanism reflecting this modification is the following:

$$A + A \underset{k_{-1}}{\overset{k_1}{\rightleftharpoons}} A^* + A \tag{5.14}$$

$$A^* \xrightarrow{k(E)} A^\ddagger \tag{5.15}$$

$$A^\ddagger \xrightarrow{k^\ddagger} \text{products} \tag{5.16}$$

After the reversible reaction of acquiring/loosing energy, the next step is energy redistribution in the emerging energised molecules A* leading to concentration of the necessary energy in the critical vibrational mode of the final step – the unimolecular reaction – whose rate coefficient $k(E)$ is energy dependent. The actual

---

[6]Oscar Knefler Rice (1903–1978) was an American physical chemist working in the field of quantum chemistry and statistical thermodynamics. Herman Carl Ramsperger (1896–1932) was also an American physical chemist studying unimolecular gas reactions. Ramsperger studied the problems connected to the Lindemann theory in case of the decay of azomethane. (It turned out later that this is a radical-propagated chain reaction!) The two of them published a paper in 1927 to propose the solution to the problem. Louis Stevenson Kassel (1905–1973) was an American chemist who published his practically equivalent solution to the problem also in 1927.

decay into products happens within one period of the critical vibration of the transition-state molecule $A^{\ddagger}$ with the associated rate coefficient $k^{\ddagger}$.

Every molecule A having an energy greater than E_0 will react; but the rate of reaction depends on the actual energy content. Following Kassel's treatment, the distribution of energy quanta in the energised molecule is random. As the energy of the molecule (and the volume) is fixed, each microstate is equally probable – according to the corresponding microcanonical ensemble. As stated before, vibrational modes are responsible for the reaction, thus we should calculate the probability that from n vibrational quanta residing within the s modes of the molecule, the necessary m are concentrated on the critical vibrational mode leading to reaction. This problem can be traced back to the distribution of n pebbles in s boxes, or equivalently – using combinatorial terms – to the permutation with repetition of n pebbles (elements of first kind) and $s-1$ separators (elements of second kind). The number of possible arrangements can be given with the following combinatorial identity:

$$P_{(n,s-1)}^{n+s-1} = \frac{(n+s-1)!}{n!(s-1)!}. \qquad (5.17)$$

This is the number of microstates of a molecule containing vibrational energy $nh\nu$. In case when m quanta necessary for reaction are localised in the critical vibration, the remaining $n-m$ quanta should be distributed among s vibrational modes. The number of possible arrangements is again given by a permutation with repetition:

$$P_{(n-m,s-1)}^{n-m+s-1} = \frac{(n-m+s-1)!}{(n-m)!(s-1)!}. \qquad (5.18)$$

The probability that $E_0 = mh\nu$ energy is localised in the critical vibration from the total vibrational energy $E = nh\nu$ is the ratio of these two permutations:

$$P(E) = \frac{(n-m+s-1)!}{(n-m)!(s-1)!} \frac{n!(s-1)!}{(n+s-1)!} = \frac{(n-m+s-1)!n!}{(n-m)!(n+s-1)!}. \qquad (5.19)$$

If the molecules are large enough, we can use the approximation that $n \gg s$ and $m \gg s$. Without going into the details of the approximate calculation, the result can be written as:

$$P(E) \cong \left(1 - \frac{m}{n}\right)^{s-1}. \qquad (5.20)$$

Let us make use of the equalities $E_0 = mh\nu$ and $E = nh\nu$; thus we can write:

5.2 Theories of Unimolecular Gas Reactions

$$P(E) \cong \left(1 - \frac{E_0}{E}\right)^{s-1}. \tag{5.21}$$

(This is identical to the result obtained by Rice and Ramsperger using classical statistical physics.)

Using the plausible assumption that the vibrational energy randomly distributes after each vibrational period, the ratio of the transition-state molecules $A^{\ddagger}$ to the energetic molecules A^* always reflects the above probability; consequently, we can also write:

$$\frac{k(E)}{k^{\ddagger}} \cong \left(1 - \frac{E_0}{E}\right)^{s-1}, \tag{5.22}$$

from which we can express the energy-dependent rate coefficient:

$$k(E) \cong k^{\ddagger}\left(1 - \frac{E_0}{E}\right)^{s-1}. \tag{5.23}$$

An interesting conclusion from this result is that, at high enough energy where $E_0 \ll E$, $k(E) \cong k^{\ddagger}$. Accordingly, a highly enough energised molecule decays within one vibrational period; there is no need for further IVR.

However, it is clear that this is only the microcanonical rate coefficient as it refers to *constant energy*. We should integrate this expression to get the expectation value of $k(E)$ at *constant temperature*, based on the Boltzmann distribution. This expectation value is then the isothermal rate coefficient k_2 of the Lindemann-Hinshelwood mechanism according to the RRK theory.

Summing up the RRK theory we can state the following. The theory includes the energy-dependent rate coefficient k_1 for the activation via collisions according to the Lindemann-Hinshelwood theory; along with the energy-dependent rate coefficient k_2 for the subsequent unimolecular reaction, taking into account the structure of the molecule as well. Though it gives a fairly good estimate for rate coefficients, it still contains quite crude approximations. One of them is the assumption that the frequency of all the *s* vibrational modes is identical. (This is the *resonance condition*; without it, there is no transition of vibrational energy from one (harmonic) oscillator to another.) The other approximation is that the total vibrational energy $nh\nu$ of the energised molecule, as well as the energy required in the critical vibration $mh\nu$ is assumed to be much greater than the ground state vibrational energy $sh\nu$ of the molecule.

5.2.4 RRKM Theory

At the time when Marcus further developed the RRK theory as a young co-worker of Rice, transition-state theory was already known to most of the chemists, and quantum-chemical methods were developed that weren't known at the time of the conception of the RRK theory. Based on these achievements, Marcus[7] had introduced several improvements to the calculation of the microcanonical rate coefficient $k(E)$. These improvements included the following:

- He considered molecular vibrations with their proper (different) frequencies. He has proved that – due to the anharmonicity of vibrations – energy transfer is possible within oscillators of different frequencies. (Anharmonic vibrations can be interpreted in terms of a series expansion which contains several higher harmonics and their linear combination as well. Among these frequencies, there are some that have approximately the same value in case of different vibrational modes.)
- He determined the microcanonical vibrational density of states (the number of microstates) using correct quantum-mechanical energy calculations. He also took into account that the zero-point energy (the energy of the vibrational ground state) cannot change during internal vibrational redistribution. (The lowest possible vibrational energy is that of the vibrational ground state, which is populated already at 0 K temperature.)
- He accounted for the effect of the rotational energy for the rate of reaction. He realised that – for both energy and momentum (the latter associated to the rotational quantum number J) of the molecule is conserved during the transformation of the molecules – the increasing bond distance during the formation of the more loosely bound transition state (*i.e.* the increase in the momentum of inertia) results in a diminished rotational energy; thus, part of it transforms into vibrational energy, increasing the rate of the reaction.

Using actually available advanced quantum chemical calculations and the modifications introduced by Marcus, very good quality rate coefficients k_{uni} can be obtained, in a good agreement with experimental data. Introduction of more advanced quantum chemical methods do not change the basic concepts of the RRKM theory; thus, these new methods do not imply any change in the name of the RRKM theory.

Problems

1. The bond dissociation energy of NO_2 in gas phase is 300.5 kJ/mol. NO_2 can undergo photodissociation according to the process $NO_2 + h\nu \rightarrow NO + O$. Calculate the critical wavelength of light that can activate an NO_2 gas molecule to dissociate.

---

[7]Rudolph Arthur Marcus (1923–) is a physical chemist of Canadian origin who later became a citizen of the United States. He has received the 1992 Nobel Prize in Chemistry for his contributions to the theory of electron transfer reactions in chemical systems – by that time already named after him. He published his results concerning the theory nowadays known as RRKM in his publications in 1951 and 1952.

5.2 Theories of Unimolecular Gas Reactions

Solution: To activate the NO_2 molecule, the exciting photon should have an energy which is at least 300.5 kJ/mol photon, or greater. For one photon, this energy is 300,500 J/6.022 × 10^{23} = 4.99 × 10^{-19} J.

The energy of a photon is $h\nu = \frac{hc}{\lambda}$, where h is Planck's constant, ν is frequency, c is the velocity of light, and λ is the wavelength of the photon. From this, we can express the wavelength: $\lambda = \frac{hc}{h\nu}$.

The result is $\lambda = \frac{6.626 \times 10^{-34} \text{ J s} \cdot 2.998 \times 10^{8} \text{ ms}^{-1}}{4.99 \times 10^{-34} \text{ J}} = 3.981 \times 10^{-7}$ m.

Thus, to dissociate a gas phase NO_2 molecule, 398.1 nm or shorter wavelength light is necessary. (This is just within the lower wavelength edge of the visible spectrum and available in the sunlight at the surface of the earth.)

2. The rate coefficient k_{uni} of a unimolecular gas reaction at a given temperature and pressure is 15% of the high-pressure limit k_∞. If the reaction follows the Lindemann mechanism, what will be the value of k_{uni} at a pressure three times higher if the reaction starts from the pure reactant? Let us suppose that the reaction mixture can be described by the ideal gas law.

Solution: Let us start from Eq. (5.9):

$$k_{uni} = \frac{k_\infty}{1 + \frac{k_\infty}{k_1[A]}}.$$

First we should solve the equation $0,15\ k_\infty = \frac{k_\infty}{1+\frac{k_\infty}{k_1[A]}}$ to get the value of $k_1[A]$. The solution is

$$k_1[A] = \frac{0.15}{1 - 0.15}\ k_\infty = \frac{0.15}{0.85}\ k_\infty.$$

The collisional activation rate coefficient k_1 is independent of pressure, and the ideal gas law renders the concentration [A] proportional to the pressure; thus three times higher pressure corresponds to $k_1[A] = \frac{3 \cdot 0.15}{0.85}\ k_\infty$. Substituting this into Eq. (5.9), we get

$$k_{uni} = \frac{k_\infty}{1 + \frac{k_\infty}{\frac{3\cdot 0.15}{0.85}k_\infty}} = \frac{1}{1+\frac{1}{\frac{3\cdot 0.15}{0.85}}}\ k_\infty = \frac{1}{1+\frac{0.85}{3\cdot 0.15}}\ k_\infty = 0.346\ k_\infty.$$

Thus, the unimolecular rate coefficient k_1 is 34.6% of the high-pressure limit k_∞ at three times higher pressure than the original one.

3. Let us consider the following unimolecular gas reaction:

$$NH_2OH \rightarrow NH_2 + OH$$

(a) What is the unimolecular rate coefficient k_{uni} at 298 K (close to 25 °C) temperature and 1 bar pressure according to the Lindemann model if the limiting reaction rate coefficients at zero and infinite pressures can be given with the following parameters of the extended Arrhenius equation:

$$a_1 = 1.62 \times 10^{-1} \text{ cm}^3 \text{ molecule}^{-1} \text{ s}^{-1}, n_1 = -5.96, E_1 = 2.79 \times 10^5 \text{ J/mol}$$

$$a_\infty = 8.03 \times 10^{16} \text{ s}^{-1}, n_\infty = -1.31, E_\infty = 2.68 \times 10^5 \text{ J/mol}.$$

(The exponent n is for a factor calculated as $(T/298 \text{ K})^n$.)

(b) What is the change of the reaction rate at the same temperature when increasing the pressure by a factor of 2, 3 and 11?

Solution: (a) To calculate the rate coefficients k_1 and k_∞, we should use the extended Arrhenius equation in the form: $k = a\,(T/298 \text{ K})^n\, e^{-\frac{E}{RT}}$. As we make the calculation for 298 K, the temperature factor becomes always unit.

Accordingly:

$$k_1 = 1.62 \times 10^{-1} \text{cm}^3 \text{ molecule}^{-1} \text{s}^{-1} \cdot e^{-\frac{2.79 \times 10^5 \text{ Jmol}^{-1}}{8.315 \text{ J K}^{-1} \text{ mol}^{-1} \cdot 298 \text{ K}}}$$

$$= 2{,}039 \times 10^{-50} \text{cm}^3 \text{ molecule}^{-1} \text{s}^{-1}$$

and

$$k_\infty = 8.03 \times 10^{16} \text{s}^{-1} \cdot e^{-\frac{2.68 \times 10^5 \text{ Jmol}^{-1}}{8.315 \text{ J K}^{-1} \text{ mol}^{-1} \cdot 298 \text{ K}}} = 8{,}561 \times 10^{-31} \text{s}^{-1}.$$

To calculate k_{uni} at 1 bar pressure, we should use Eq. (5.9) with the limiting first order low pressure value of the rate coefficient, but the reported value is given as a second-order rate coefficient k_1; thus we should multiply the reported value by the number of molecules at 298 K and 1 bar – as it is in Eq. (5.9). Supposing that NH_2OH gas follows the ideal gas law under these conditions, this concentration is:

$$[A] = \frac{N_A P V}{RT} = \frac{6.022 \text{ mol}^{-1} \cdot 1 \text{cm}^3 \cdot 1 \text{ bar}}{8.315 \text{ J K}^{-1} \text{ mol}^{-1} \cdot 298 \text{ K}} = \frac{6.022 \text{ mol}^{-1} \cdot 10^{-6} \text{m}^3 \cdot 10^5 \text{ Pa}}{8.315 \text{ J K}^{-1} \text{ mol}^{-1} \cdot 298 \text{ K}}$$
$$= 2{,}430 \times 10^{19} \text{cm}^{-3}.$$

Using the above calculated values, we get

$$k_{uni} = \frac{k_\infty}{1 + \frac{k_\infty}{k_1[A]}} = \frac{8{,}561 \times 10^{-31} \text{s}^{-1}}{1 + \frac{8{,}561 \times 10^{-31} \text{s}^{-1}}{2{,}039 \times 10^{-50} \text{cm}^3 \text{s}^{-1} \cdot 2{,}430 \times 10^{19} \text{cm}^{-3}}}$$

As a result, we get $k_{uni} = 3.139 \times 10^{-31} \text{ s}^{-1}$ (related to concentrations in cm^{-3} units).

(b) To calculate the change of the rate coefficient due to increasing the pressure, we start from the same relation: $k_{\text{uni}} = \frac{k_\infty}{1+\frac{k_\infty}{k_1[A]}}$. As the collisional activation rate coefficient k_1 is independent of pressure, and the ideal gas law renders the concentration [A] proportional to the pressure, x times higher pressure corresponds to $k_{\text{uni}} = \frac{k_\infty}{1+\frac{k_\infty}{x\,k_1[A]}}$. Applying this relation to $x = 2$, 3 and 11 times the original 1 bar; we get 4.139×10^{-31} s^{-1}, 5.433×10^{-31} s^{-1} and 7.399×10^{-31} s^{-1}, respectively.

Further Reading

1. Pilling MJ, Seakins PW (1995) Reaction kinetics. Oxford University Press, Oxford
2. de Paula J, Atkins PW (2014) Physical chemistry. 10th edn. Oxford University Press, Oxford
3. Silbey LJ, Alberty RA, Moungi GB (2004) Physical chemistry. 4th edn. Wiley, New York
4. Steinfeld JI, Francisco JS, Hase WL (1998) Chemical kinetics and dynamics. 2nd edn. Prentice Hall, Englewood Cliffs
5. Forst W (1973) Theory of unimolecular reactions. Academic, New York
6. Keszei E (2012) Chemical thermodynamics – an introduction. Springer, Berlin

Chapter 6
Catalysts and Catalytic Reactions

If there were no catalytic reactions, there would be no life; *enzymes* – *biocatalysts* found in living organisms – allow for the great number of reactions under physiological conditions necessary for normal functions of these organisms. Current theories concerning the origin of life assume that claylike minerals catalysed the formation of first highly organised molecular structures necessary to form the first primitive living cells. To prepare food that can be stored for longer times, mankind made use of catalytic reactions since thousands of years; bread, yoghurt, kefir, wine and beer are prepared by microorganisms doing their job by enzyme-catalysed reactions.

However, during the development of chemistry, it was realised quite late that there are special reactants that are now called catalysts[1] – a name that has been proposed by Berzelius.[2] Satisfactory understanding of the mechanism of their catalytic effect was only possible after molecular details of these reactions have been elucidated.

According to our actual knowledge we can say that catalysts exert their accelerating effect on reactions by acting repeatedly in a closed-cycle mechanism. However, in contrast to chain reactions – which need an initialisation reaction first and have termination steps as well – catalysts are already present in the reaction mixture prior to reaction[3] and there is no terminating step that would end the catalytic activity. In a typical catalytic mechanism, the catalyst forms an intermediate with one of the reactants, which opens up a reaction path for a quick transformation, after which the catalyst is released and can form another reactive intermediate.

---

[1]The word catalysis is derived from the Greek verb κατάλυω, meaning to destroy or disassemble. The reason for giving this name is that the first known catalysts accelerated decay reactions, or made them possible at all.

[2]Jöns Jacob Berzelius (1779–1848) was a Swedish chemist. He had an important role in the development of modern chemistry; *e. g.* he discovered electrolysis, and he proposed to use modern symbols to identify elements.

[3]Autocatalysis is an exception which will be discussed later.

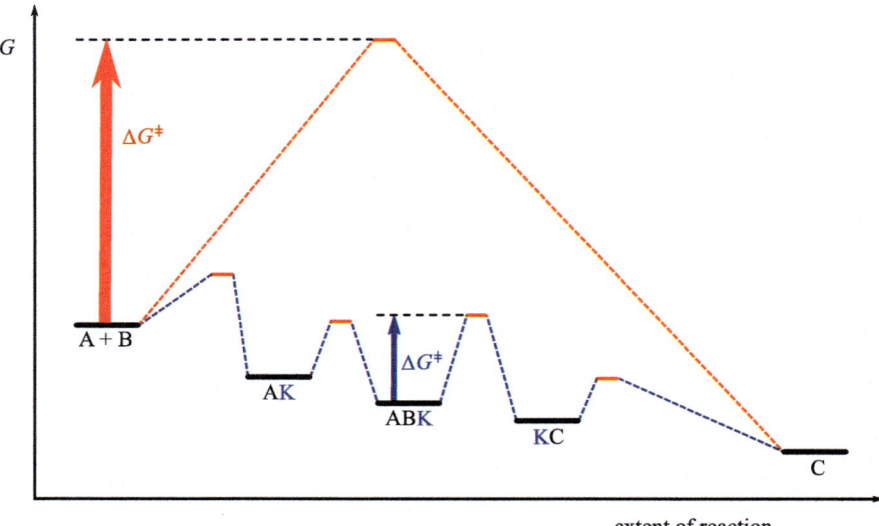

Fig. 6.1 Schematic free energy diagram of the reaction A + B → C without a catalyst (red dashed lines) and in case of the catalytic cycle operating with the catalyst K (blue dashed line). It is clear that the highest activation energy step in case of the catalytic reaction (ABK → KC; blue arrow) is much lower than the activation energy of the non-catalytic reaction (red arrow)

The mechanism shown below is an example for an alternative catalytic reaction path of a bimolecular reaction A + B → C.

$$A + K \rightarrow AK \tag{6.1}$$

$$AK + B \rightarrow ABK \tag{6.2}$$

$$ABK \rightarrow KC \tag{6.3}$$

$$KC \rightarrow K + C \tag{6.4}$$

We can see that the catalyst K reacts in the first step (6.1) with the molecule A but in the final step (6.4) of the mechanism it becomes free again, ready to react with another molecule A. Adding reactions (6.1)–(6.4) we get back the stoichiometric equation of the non-catalysed reaction A + B → C. The explanation of the effect of the catalyst to increase the rate of the reaction can be visualised with a schematic free energy diagram of the mechanism (6.1)–(6.4).

We can see from Fig. 6.1 that the contribution of the catalyst to the reaction is to open up a reaction path with considerably lower activation energy than the one without the catalyst. Recalling the Arrhenius equation (2.51) it is clear that the rate of the catalysed reaction is increased by the factor of the exponential function of the difference of the two activation Gibbs potentials. If the activation Gibbs potential of the non-catalysed reaction is so high that the available thermal energy is not

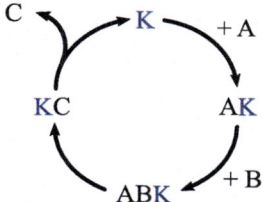

Fig. 6.2 Catalytic cycle of the reaction A + B → C with the participation of the catalyst K. This reaction proceeds over and over again until reactants A and B are present in the reaction mixture, producing component C (For the sake of simplicity, we have written only irreversible reactions in the catalytic cycle. However, it is possible that components AK and ABK also dissociate into K + A and AK + B, prior to further reaction in the mechanism shown)

sufficient to overcome the activation barrier, the reaction would not occur at all without the catalyst; while with the catalyst, the lower activation Gibbs potential is available from the thermal energy of the molecules. The cyclic nature of the catalytic mechanism (6.1)–(6.4) can be visualised in a scheme shown in Fig. 6.2.

We can classify catalysts in three groups. *Homogeneous catalysts* are in the same phase as reactants and they are miscible with them. *Heterogeneous catalysts* are in a different phase than reactants and products, immiscible with the phase containing the reaction mixture. *Biocatalysts* are typically in the same phase as the reaction mixture and are also miscible with it. However, these are large protein molecules reactants are binding to (into the *binding pocket* or *active site*). This binding reaction is called *docking*. The conformation of the docked reactant changes in a way that largely facilitates the catalysed reaction. After the reaction, the product is detached from the active site and the enzyme can enter another catalytic cycle.

The role of the three types of catalysts is quite similar. They all take part in the reaction as reactant; thus, this reaction can formally be written in the reaction mechanism. In many instances, the partial order of the reaction with respect to the homogeneous catalyst is 1; in this case, the rate of the catalysed reaction is proportional to the concentration of the catalyst.

An interesting property of catalysts is that they do not alter the equilibrium of the reaction (*i.e.* the equilibrium concentration of the reactants and products), but they typically increase the rate of reaction in both directions. As we can see, this effect is due to the decrease of the activation Gibbs potential (at constant pressure and temperature) – as shown schematically in Fig. 6.1. Taking into account the identity $\Delta G^{\ddagger} = \Delta H^{\ddagger} - T \Delta S^{\ddagger}$, the catalyst can either decrease the activation enthalpy $\Delta H^{\ddagger}$, or increase the activation entropy $\Delta S^{\ddagger}$, or both.

Strictly speaking we call catalyst only the substance that *increases* the rate of a reaction, or makes possible a reaction that would not happen without the catalyst. Substances that *decrease* the rate of reactions – or stop them completely – are called *inhibitors*.[4] In case of simple (pseudo-first order) effects, the influence on the

---

[4]The word inhibitor is derived from the Latin verb *inhibeo*, meaning to hinder, or prevent.

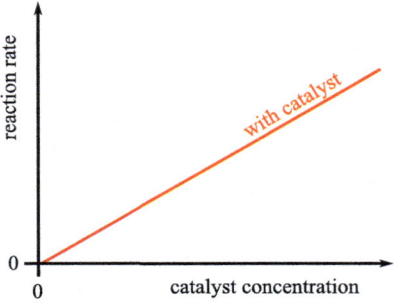 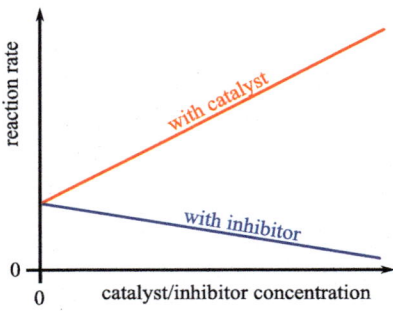

Fig. 6.3 Rate of catalysed and inhibited reactions as a function of catalyst/inhibitor concentration in case of simple (first order in catalyst/inhibitor) reactions. Left panel: the reaction does not go without catalyst. Right panel: the reaction goes also without catalyst with a moderate rate

reaction rate by a homogeneous catalyst and an inhibitor as a function of its concentration is shown in Fig. 6.3.

An inhibitor hinders or fully prevents a reaction by removing an intermediate from the reaction which is the key element in a low activation energy reaction path. If this 'side reaction' is much faster than the formation of the product without inhibitor, then the reaction can be completely stalled. Typical inhibitors are radical scavengers that capture an intermediate radical species.

In case of homogeneous catalysis, the suitable method to study the reaction is to explore the mechanism treating the catalyst as a reactant and also as a product. This kind of mechanism shows similarities with that of unbranched chain reactions. In case of heterogeneous catalysis – discussed in the next section – component transfer between different phases should also be taken into account.

6.1 Heterogeneous Catalysis

In a heterogeneous catalytic reaction, reactants and catalyst are in different phases, and the reaction itself proceeds at the phase boundary. Possible couples are solid–solid, liquid–liquid, solid–liquid, solid–gas and liquid–gas interfaces. Solid–solid heterogeneous reactions are quite frequent in the Earth's crust, but diffusion in solids is so slow that these reactions take geological time scales to happen. In case of an interesting and industrially relevant liquid–liquid heterogeneous catalysis method the catalyst dissolves only in one phase which – within a certain temperature range – is immiscible with the other phase containing reactants. By changing – typically, increasing – temperature, the two phases mix completely. After mixing, a homogeneous catalytic reaction takes place in the single liquid phase. At the completion of the reaction, temperature is decreased and phase separation occurs. Products remain in the phase where originally the reactants were, while the catalysts are contained in the newly separating phase. Typical examples are aqueous–organic, organic–

organic, aqueous–fluorous and organic–fluorous biphasic reactions. However, these reactions proceed via homogeneous catalysis; phase separation only largely simplifies extraction of the catalyst from the product mixture, thus enabling its easy reuse.

Solid–liquid heterogeneous catalytic reactions are not very frequent in industrial applications; however, formation of supramolecular assemblies and complicated molecular structures are often catalysed by suitable solids facilitating their formation. (Current theories explain the formation of basic organic molecules necessary to the evolution of living matter having been catalysed by clay minerals.)

A well-known liquid–gas heterogeneous reaction is the formation of carbonic acid following the absorption of CO_2 in water. However, carbonic acid formed this way in small quantity does not return back to the gas phase; thus, it is not a heterogeneous catalytic procedure. Industrial liquid–gas heterogeneous catalytic reactions – where reactants are in the gas phase and the catalyst in the liquid phase – occur mostly in a thin liquid layer on the surface of solids. Mechanism of these reactions is very similar to that of solid–gas heterogeneous reactions; transformation of reactants into products happens on the surface of the liquid adsorbed on the solid surface. There are large-scale industrial electrochemical reactions where reactants are transformed into products on the surface of one of the electrodes during electrolysis. Among these processes, there are also *electrocatalytic* reactions, which are treated within the framework of electrochemistry.

By far the most used heterogeneous catalytic reactions in industry are gas reactions on the surface of solid catalysts; thus, we shall describe some details of these reactions. The most important difference with respect to the three-dimensional (*bulk*) reactions is that the description of surface reaction is in terms of (two-dimensional) *surface concentrations* – as they only take place on the surface of the solid. Its dimension is amount of substance/surface area, the SI unit being mol/m^2.

Transformation of a fluid phase reactant on a surface is always a composite reaction where processes should happen in the following order:

1. Diffusion of reactants onto the surface
2. Adsorption (and eventual desorption) of the reactants.
3. Surface diffusion to an active site on the surface, or diffusion of reactants on the surface to encounter.
4. The surface reaction itself.
5. Desorption of the products from the surface.
6. Diffusion of the products into the bulk fluid phase.

In case of a unimolecular surface reaction when reactants are adsorbed only at active sites on the surface, step 3 is not needed. In step 5, re-adsorption of the products is not necessary to take into account as their concentration in the fluid phase is typically much smaller than on the surface.

Let us first consider a unimolecular reaction during which reactant A is transformed into some product(s) on the catalytic surface. Let us denote the free surface site where the reaction can occur by S. (This is a special molecule or atom at the surface.) We can write the mechanism of this reaction the following way:

$$A + S \underset{\text{desorption},k_{-1}}{\overset{\text{adsorption},k_1}{\rightleftarrows}} AS \tag{6.5}$$

$$AS \xrightarrow{k'_2} \text{products} \tag{6.6}$$

When describing surface reactions, it is practical to define a dimensionless quantity, the *surface coverage*, defined as the number of adsorbed molecules on the surface divided by the number of molecules in a fully covering monolayer on that surface. With the help of this quantity, we can describe the rate of reaction using the following notation:

c – overall concentration of active centres on the surface.
θ – fraction of active centres on the surface covered by molecules A.
$1 - \theta$ – fraction of the uncovered active centres on the surface
$c_{AS} = c\,\theta$ – surface concentration of A adsorbed at active centres.
$c_S = c\,(1 - \theta)$ – surface concentration of free active centres (not covered).

Accordingly, the surface coverage θ can be expressed as c_{AS}/c. The rate of reaction (6.6) is also given as a surface reaction rate; therefore, we can write

$$\frac{d[\text{products}]}{dt} = r = k'_2 c_{AS} = k'_2 c \theta \tag{6.7}$$

Let us introduce the simplified notation $k'_2 c = k_2$ based on the constant value of active surface centres (independent from the quantity of molecules A adsorbed).[5] Using this notation, the rate of the surface reaction can be written as $r = k_2\,\theta$; which is the usual form of the two-dimensional first-order rate equation. However, there is no possibility to measure the surface coverage θ as – even in the case of a stationary state – it is determined by two processes: the formation and the decomposition of the surface complex AS; and there is no method yet to determine θ when a reaction proceeds. In case of a stationary reaction, we can write the following approximation, using c_A as the *bulk* (three-dimensional) concentration of the reactant A in the gas phase:

$$k_1 c_A (1 - \theta) \cong k_{-1} \theta + k_2 \theta \tag{6.8}$$

The left-hand side is the rate of formation of the surface complex AS; proportional to the gas-phase concentration c_A and to the free active surface fraction $1 - \theta$. The first

---

[5]The overall concentration of active surface centres does not change if there is no *catalyst poisoning*. This would happen if one of the reactive mixture components deactivated some active centres as a result of chemical change.

6.1 Heterogeneous Catalysis

term on the right-hand side is the rate of active centre S becoming free while molecule A desorbs; the second term being the rate of transformation of the complex AS, along with the desorption of the product(s). The last two terms are proportional to the surface coverage θ (the ratio of the surface concentration of the complex AS to that of the overall concentration of active centres S). Rearranging Eq. (6.8) we can express the approximate value of θ becoming constant during the stationary reaction:

$$\theta \cong \frac{k_1 c_A}{k_1 c_A + k_{-1} + k_2}, \tag{5.9}$$

from which we can have the following result (making use of the relation $r = k_2 \theta$) for the rate of the surface reaction:

$$r = \frac{k_1 k_2 c_A}{k_1 c_A + k_{-1} + k_2}. \tag{6.10}$$

(We have ceased to show the 'approximately equal' $\cong$ symbol.) In this expression there is only the gas phase concentration c_A of the component A, in addition to rate coefficients. Note that care should be taken to choose rate coefficients k_1, k_{-1} and k_2 (and their units) so that Eqs. (6.6)–(6.10) be consistent concerning their dimensions and units.

Let us explore chemically relevant limiting cases of Eq. (6.10). In case when the surface reaction is much quicker than adsorption and desorption; that is, $k_2 \gg k_1 c_A + k_{-1}$, then

$$r \cong \frac{k_1 k_2 c_A}{k_2} \cong k_1 c_A \tag{6.11}$$

and

$$\theta \cong \frac{k_1 c_A}{k_2}. \tag{6.12}$$

Thus, the catalytic reaction is of first-order with respect to the bulk concentration while the surface coverage is very low; as $k_2 \gg k_1 c_A$ implies $\theta \ll 1$. This can be understood considering that adsorption is slow compared to surface reaction and subsequent desorption. Examples for this case are decomposition of N_2O to its elements on gold surface, or decomposition of HI to its elements on platinum surface.

In case when the surface reaction is much slower than adsorption and desorption; that is, $k_2 \ll k_1 c_A$ and $k_2 \ll k_{-1}$, then

$$r \cong \frac{k_1 k_2 c_A}{k_1 c_A + k_{-1}} = \frac{k_2 \frac{k_1}{k_{-1}} c_A}{\frac{k_1}{k_{-1}} c_A + 1}. \tag{6.13}$$

Let us consider $\frac{k_1}{k_{-1}} = K$ as the adsorption equilibrium constant of A and substitute into the above expression:

$$r \cong \frac{k_2 K c_A}{K c_A + 1} \tag{6.14}$$

and

$$\theta \cong \frac{K c_A}{K c_A + 1}. \tag{6.15}$$

Obviously, the dependence of the reaction rate r on the bulk concentration c_A is more complicated in this limiting case. However, the expression of the surface coverage θ is identical to the *Langmuir-Hinshelwood* two-dimensional equation of state which describes the adsorption equilibrium. This result is not surprising: the surface reaction is much slower than adsorption and desorption, thus the equilibrium value of θ can always be maintained.

Two limiting cases of Eqs. (6.14)–(6.15) are also worth discussing. If the gas phase bulk concentration c_A is so small that $K c_A \ll 1$, then $r \cong k_2 K c_A$, which means that the gas phase reaction is of first-order with the rate coefficient $k_2 K$. Surface coverage within these conditions is proportional to the bulk concentration c_A: $\theta \cong K c_A$. In the opposite case when $K c_A \gg 1$, then $r \cong k_2$, which means that the reaction is of zero-order and the coverage is almost complete: $\theta \cong 1$. Examples for this case are decomposition of HI, or NH_3 on gold surface.

Observing Eqs. (6.10) and (6.14) we can see that (apart from the difference of the constants a and b) both have the form $r = \frac{a c_A}{b c_A + d}$. Thus, if we trace the reaction rate r as a function of the bulk concentration c_A, we get the curve shown in Fig. 6.4. This curve reveals that at very low bulk concentration the reaction is of first-order, while in case of very high bulk concentration, it is of zero-order.

If the catalysed surface reaction is bimolecular, we can follow the previous procedure and consider a stationary reaction; the only difference is the somewhat more complicated mechanism. Molecules of the two reactants (let them denote by A and B) should first be adsorbed, then encounter by diffusion on the surface, react with the assistance of the catalytic site to give the product, which then should also be desorbed from the surface. Without going into details of this mechanism, we show here only the surface reaction and the formula for the reaction rate. The (two-dimensional) surface reaction can be written as

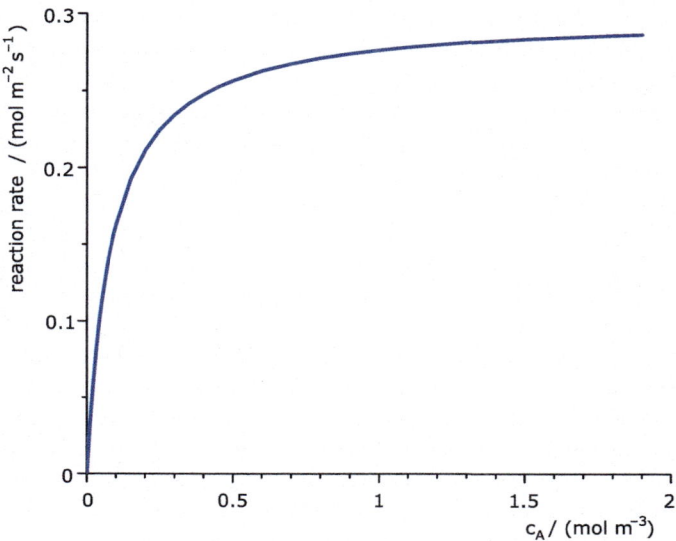

Fig. 6.4 Plot of the functions of Eqs. (6.10) and (6.14), with the common parameters $a = 16$, $b = 50$ and $d = 8$ (see text). At small concentrations (between 0 and cca. 0.1 mol m^{-3}) it is approximately linear, then it curves sharply before turning into a horizontal line at high concentrations. Along the first linear part, reaction rate is proportional to the concentration; that is, the reaction is of first-order. After it becomes (almost) horizontal, the reaction rate is (approximately) independent of concentration; thus, the reaction is of zero-order

$$\text{AS} + \text{BS} \xrightarrow{k'} \text{products}. \tag{5.16}$$

The rate of this reaction is as follows:

$$\frac{d[\text{products}]}{dt} = r = k\theta_A \theta_B. \tag{6.17}$$

Here, θ_A is the fraction of the surface covered by A and θ_B its fraction covered by B. Using the same stationary approximation and the limiting case of a chemical reaction much slower than adsorption and desorption, the result for the rate of reaction in terms of the bulk concentration is the following:

$$r \simeq \frac{kK_A K_B c_A c_B}{K_A c_A + K_B c_B + 1} \tag{6.18}$$

We can see that the structure of this expression is quite similar to the one obtained for a unimolecular reaction; the difference being that adsorption equilibrium constants of both components are involved, and the numerator is proportional to the product of the two bulk concentrations. Thus, this expression is also related to the Langmuir-Hinshelwood equation of state when two components get adsorbed on the surface.

Limiting cases of this rate expression can also be calculated similarly as before; the difference is only that we need more conditions in the approximations and the results are also more complex.

It is worth noting that gas-phase molar concentrations are easy to convert to partial pressure. As pressure is convenient to control, it is used in industrial practice instead of molar concentration.

6.2 Enzyme Catalysis

As already mentioned at the beginning of this chapter, biocatalysts are usually big – typically of several thousand dalton[6] molar mass – protein molecules that selectively catalyse physiologically important reactions in living organisms. Selectivity in this respect means that they only catalyse one single reaction on a specific reaction path; that is, into a specific product. The reactant of an enzyme catalysed reaction is called *substrate*. The structure of enzyme involves an *active site* (also called *binding pocket*) where the substrate gets bound, which in turn opens up a low activation energy reaction path. (This is the reason to consider enzyme reactions similar to heterogeneous catalytic reactions; however, in most of the cases the enzyme is present in the same (liquid) phase as its substrate.)

Many reactions of a substrate S catalysed by an enzyme E can be described by the following simple mechanism:

$$E + S \underset{k_{-1}}{\overset{k_1}{\rightleftarrows}} ES \qquad (6.19)$$

$$ES \overset{k_2}{\rightarrow} E + \text{products} \qquad (6.20)$$

This is called the *Michaelis-Menten mechanism*.[7] In principle, the second step may be reversible as well; however, under the conditions of functioning living organisms it is mostly irreversible, proceeding towards the products. (Note that this mechanism is formally quite similar to that of a unimolecular heterogeneous catalytic process

---

[6]Dalton is the name of the unit of molar mass; 1 Da = 1 g/mol. (It is a non-SI unit but frequently used in biochemistry.)

[7]Leonor Michaelis (1875–1949) was a German medical doctor and biochemist who had worked until 1922 in Berlin; then – after a brief period spent in Japan – he worked in the United States. Maud Leonora Menten (1879–1960) was a medical researcher; the first women in Canada who graduated as a medical doctor. Between 1905 and 1917 she worked with Michaelis, spending several years in Berlin. They authored a paper in 1913 summarising their studies of enzyme catalysed reactions, along with a theoretical model which has been named later as Michaelis-Menten mechanism. Their model calculation was based on a fast pre-equilibrium, but their results are equivalent to those of Briggs and Haldane who developed the model based on QSSA for the ES complex in 1925 – the same as shown here.

6.2 Enzyme Catalysis

according to Eqs. (6.5)–(6.6), as well as a thermally activated gas-phase unimolecular reaction according to the Lindemann mechanism (5.1)–(5.2); thus, we can expect similar result for the rate of reaction as in those cases.) Rate equations for this mechanism can be written as

$$\frac{d[\text{ES}]}{dt} = k_1[\text{E}][\text{S}] - k_{-1}[\text{ES}] - k_2[\text{ES}] \qquad (6.21)$$

$$\frac{d[\text{S}]}{dt} = k_{-1}[\text{ES}] - k_1[\text{E}][\text{S}] \qquad (6.22)$$

$$\frac{d[\text{P}]}{dt} = k_2[\text{ES}] \qquad (6.23)$$

Here, P denotes the product(s) of the reaction. We can treat the intermediate ES complex as a quasi-stationary state (QSS) species; thus we can equate its rate of formation (left-hand side) to the rate of its decay (right-hand side):

$$k_1[\text{E}][\text{S}] \cong k_{-1}[\text{ES}] + k_2[\text{ES}]. \qquad (6.24)$$

We can express the steady-state concentration of the complex ES the usual way:

$$[\text{ES}] \cong \frac{k_1[\text{E}][\text{S}]}{k_{-1} + k_2} \qquad (6.25)$$

Substituting this into Eq. (6.23) we obtain the rate of formation of products:

$$\frac{d[\text{P}]}{dt} \cong \frac{k_1 k_2 [\text{E}][\text{S}]}{k_{-1} + k_2}. \qquad (6.26)$$

Let us introduce the following simplifying notation of the constants in Eq. (6.25):

$$\frac{1}{K_\text{M}} = \frac{k_1}{k_{-1} + k_2}. \qquad (6.27)$$

With this notation, the concentration of the enzyme-substrate complex simplifies to

$$[\text{ES}] \cong \frac{[\text{E}][\text{S}]}{K_\text{M}}. \qquad (6.28)$$

In case of experimental investigation of the rate of reaction, it is worth making use of conserved quantities based on stoichiometry: namely, that the enzyme concentration $[\text{E}]_\text{o}$ in the reaction mixture at the beginning of the reaction (at $t = 0$) is either in the form of a free enzyme E or as enzyme-substrate complex ES during the course of reaction. Thus, we can express the free enzyme concentration as $[\text{E}] = [\text{E}]_\text{o} - [\text{ES}]$, and substitute it into Eq. (6.28):

$$[ES] \cong \frac{([E]_o - [ES])[S]}{K_M}. \tag{6.29}$$

Let us rearrange this first as $K_M[ES] \cong [E]_o[S] - [ES][S]$, then as $(K_M + [S])[ES] \cong [E]_o[S]$.[8] From this we can write for the concentration of the enzyme–substrate complex:

$$[ES] \cong \frac{[E]_o[S]}{K_M + [S]}. \tag{6.30}$$

Upon substitution of this into Eq. (6.23) we obtain the rate of formation of the product (and the rate of decay of the substrate) as

$$\frac{d[P]}{dt} \cong \frac{k_2[E]_o[S]}{K_M + [S]}. \tag{6.31}$$

This is the expression of the rate of the reaction according to the quasi-steady-state approximation of the Michaelis-Menten mechanism, predicting that the rate increases with increasing concentration of the substrate – but not unlimitedly. At very large substrate concentration (when $[S] \gg K_M$), the rate goes into a finite limit:

$$\left.\frac{d[P]}{dt}\right|_\infty = r_\infty = k_2[E]_o. \tag{6.32}$$

As both k_2 and $[E]_o$ are independent of instantaneous concentration (and of reaction time), the reaction is of zero-order; its rate is independent of the concentration of the substrate. (For this reason, we can interpret the limiting rate r_∞ as a zero-order rate coefficient and denote it by k_∞, as it is often the case.) The other limiting value is close to zero substrate concentration:

$$\left.\frac{d[P]}{dt}\right|_0 = \frac{k_2}{K_M}[E]_o[S]. \tag{6.33}$$

Plotting the expression of the rate of reaction (6.31) according to the quasi-steady-state approximation of the Michaelis-Menten mechanism, it is easy to discern these two limiting cases. In Fig. 6.5, it is also shown that, at the rate of $r_\infty/2$, the

---

[8]Note that in the above calculations it is not taken into account that the substrate concentration [S] becomes lower by the amount of S bound in the enzyme-substrate complex and not transformed yet. This approximation holds only if the concentration of the substrate [S] is much higher than that of the enzyme-substrate complex [ES]. This condition is fulfilled if the initial substrate concentration $[S]_o$ is much larger than the initial enzyme concentration $[E]_o$. However, in living cells – and in enzyme kinetic experiments – this is typically the case; the excess of $[S]_o$ with respect to $[E]_o$ can be as high as a thousand.

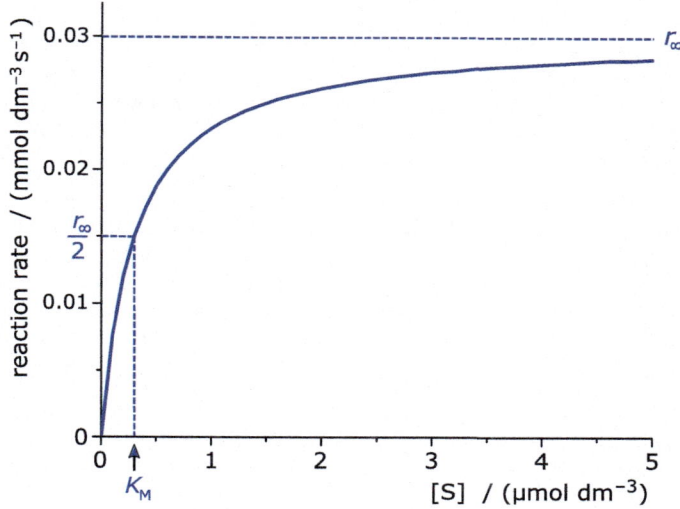

Fig. 6.5 Reaction rate of the pepsin-catalysed reaction severing proteins into peptides and amino acids, as a function of the substrate (in this case, protein). Parameters of the diagram: $[E]_o$ of the enzyme: 60 µmol/dm^3, $k_2 = 0.5$ s^{-1}, $K_M = 0.3$ µmol/dm^3. High substrate concentration limit of the rate r_∞ and the value of K_M where the rate is equal to $r_\infty/2$ are marked in the diagram. Note the similarity of the curve to that of a unimolecular heterogeneous catalytic reaction shown in Fig. 6.4

concentration is equal to the value of the Michaelis constant K_M. (This is easy to see by comparing Eqs. (6.26) and (6.32)). Observing the curve at very low and very high substrate concentrations we can see that its shape is quite similar to that of a unimolecular heterogeneous catalytic process according to Eqs. (6.5)–(6.6), as well as to that of thermally activated gas-phase unimolecular reaction according to the Lindemann mechanism (5.1)–(5.2) (see Figs. 5.1 and 6.4).

The rate of enzyme-catalysed reaction in experimental enzyme kinetic studies is written in terms of the parameters r_∞ and K_M, as these parameters can be determined experimentally by measuring the (initial) reaction rate at different substrate concentrations. We can get this expression by substituting r_∞ as $k_2[E]_o$ from Eq. (6.32) into Eq. (6.31):

$$\frac{d[P]}{dt} \cong \frac{r_\infty [S]}{K_M + [S]} = \frac{r_\infty}{\frac{K_M}{[S]} + 1}. \quad (6.34)$$

As we can see, this expression contains indeed only easily available experimental quantities.

Enzyme kinetic studies began flourishing between 1920 and 1940, when fast computers were not available. For this reason, it was common practice to transform Eq. (6.34) into linear functions that could be evaluated graphically, using a straight ruler. It is not worth any more to use these straight-line methods; however, it is useful to know that in older publications (unfortunately, sometimes even in recent ones) the evaluation of enzyme kinetic data was based on the linearised methods. The main disadvantage of this practice is that estimated parameters are biased due to linearisation (see Problem 2 of Chap. 3). However, to show this obsolete practice, we shall discuss three of these transformations into linear functions which bear the names of researchers who had proposed them first. Derivation of all these methods starts from Eq. (6.34). For the sake of simplicity, let us denote the reaction rate d[P]/dt in the equation by r.

The Lineweaver–Burk transform has the following form:

$$\frac{1}{r} = \frac{K_M}{r_\infty} \frac{1}{[S]} + \frac{1}{r_\infty}. \tag{6.35}$$

When the inverse of the reaction rate r is plotted as a function of $1/[S]$, the slope of the line is K_M/r_∞ and its intercept is $1/r_\infty$.

The Eadie–Hofstee transform has the following form:

$$r = -K_M \frac{r}{[S]} + r_\infty. \tag{6.36}$$

As we can see, in this case the reaction rate r plotted as a function of $r/[S]$ gives a line; its slope is $-K_M$ and its intercept is r_∞.

The Hanes–Woolf transform has the following form:

$$\frac{[S]}{r} = \frac{1}{r_\infty}[S] + \frac{K_M}{r_\infty}. \tag{6.37}$$

In this case the ratio $[S]/r$ as a function of the substrate concentration $[S]$ is linear, with the slope of the inverse of high-concentration rate limit $\frac{1}{r_\infty}$ and the intercept K_M/r_∞.

When finding enzyme kinetic data in literature, in case of linearised estimations it is worth checking if error propagation has been taken into account by the authors. If this was the case, parameters obtained are less biased than without error propagation. If kinetic parameters have been estimated using nonlinear estimation methods based on the model function of Eq. (6.34), then they are unbiased and the estimated standard deviations of the parameters r_∞ and K_M are also reliable.

Since the beginning of the twenty-first century there exist quite a few software applications developed specifically for treating enzyme kinetic data. These applications typically contain various enzyme kinetic models embedded; the user only has to choose the appropriate one and the relevant system of differential equations is generated by the application. Its solution is performed using numerical integration as described in Sect. 4.9; and nonlinear parameter estimation is also carried out, along with a detailed statistical inference. Another advantage of using this capacity of nonlinear estimation is that we need to have less measured data to get reliable results than in case of the linearised methods.

It is worth noting that the actual value of r_∞ also depends on the concentration of the enzyme – as it is the product $k_2[E]_0$. This is the reason to give as a parameter characteristic of the enzyme – in addition to the Michaelis constant K_M – the rate coefficient k_2 which is independent of the enzyme concentration. This parameter is called *catalytic constant* and is usually denoted by the symbol k_{cat}.

Mechanism (6.19)–(6.20) is quite simple; but many enzyme reactions are much more complex. Without going into details, we mention here that even the simple Michaelis-Menten mechanism can proceed with two kinds of *inhibitions*: *competitive* and *non-competitive*. There are also numerous enzyme-catalysed reactions where the catalyst is not one single enzyme molecule but a few so-called *subunits* associated, having multiple binding sites for the substrate. These reactions can be more or less *cooperative*: if, after the first substrate is bound to an active site, subsequent substrates can be bound more easily. This is called *positive cooperativity*. The opposite case also occurs, when the first bound substrates inhibit to some extent subsequent substrates to bind. This is called *negative cooperativity*. These more elaborate mechanisms open up the possibility for *enzyme regulation*, which means that the actual state of the living organism (usually the actual concentration of some physiologically important component) can largely influence the activity of an enzyme. These complex mechanisms are discussed in detail in textbooks of biochemistry.

6.3 Autocatalysis, Autoinhibition and Nonlinear Chemical Processes

An interesting type of catalytic reactions is the one when the catalyst is not contained in the reaction mixture at the beginning but is produced during the course of reaction, its concentration increasing with the advancement of the reaction. As the reaction 'itself' produces the catalyst, this process is called *autocatalysis* and the relevant component is called *autocatalyst*.[9] While the concentration of the autocatalyst increases during the course of reaction, the reaction is continuously accelerated. There are also cases when an inhibitor is produced during the course of reaction, resulting in a continuous slowdown of the process. In a similar manner, those processes are called *autoinhibition*, and the relevant component is called *autoinhibitor*. Explained in terms used in control theory, autocatalysis is a positive (closed loop) feedback while autoinhibition is a negative (closed loop) feedback concerning the rate of reaction. (Enzyme regulation mentioned at the end of the previous section can also work based on these principles.)

---

[9]The word autocatalyst does not mean an automotive catalytic converter; it refers to the mechanism in which the reaction itself produces its catalyst. This name is coined from two Ancient Greek words: the pronoun αὐτός (itself) and the verb κατάλυω (destroy, abolish, dissolve). See also footnote 1 on page 133.

Both autocatalyst and autoinhibitor are typically absent from the reaction mixture at the start of reaction, which facilitates the recognition of the associated behaviour. If the rate of the overall reaction has an order (*i.e.* the rate equation is a homogeneous nth order differential equation), or if we can arrange for a pseudo-order with flooding, than the rate can be measured without feedback at the very beginning of the reaction. We can measure the initial rate of the reaction using different initial concentrations of the reactants, and determine the order n_c of the reaction. In an alternative measurement we can run the reaction starting from a suitable initial concentration, and determine the rate of reaction at several successive time instants. From these data, we can also determine a reaction order n_t. (The symbol n_c refers to a purely concentration-dependent case while n_t refers to a time-dependent case.)

If there is no feedback during the course of reaction, these two orders should be identical. In case $n_t > n_c$ we can conclude that the reaction gets slower when proceeding than we could expect from the order n_c; consequently, there should be autoinhibition. In the contrary, if $n_t < n_c$, the reaction gets faster when proceeding than we could expect from the order n_c, which indicates autocatalysis. (Changes in the rate during the course of reaction depending on the order are discussed in Sect. 3.1.)

However, as it is emphasised when discussing composite reactions, rate equations in most cases are not following mass action kinetics, thus they do not have any order. In such cases autocatalysis or autoinhibition is more complicated to recognise; it is only possible with thorough study of the reaction mechanism.

6.3.1 *A Simple Autocatalytic Reaction*

The simplest autocatalytic reaction is the transformation of a substance A into B, catalysed by the product B. Though we can write the overall reaction as A $\rightarrow$ B, but the rate equation should reflect that the reactant A can only react to give the product B if it can react with the catalyst B. Therefore, it is more appropriate to write the reaction in the following form:

$$A + B \rightarrow 2B. \tag{6.38}$$

Supposing the validity of mass action kinetics[10] we can write the following rate equations:

---

[10]Concerning its stoichiometry, this is indeed the simplest autocatalytic reaction; however, we do not know any *elementary* reaction like this. Thus, this kind of reaction is a composite one, but a mass-action kinetic rate law is often applicable anyway.

6.3 Autocatalysis, Autoinhibition and Nonlinear Chemical Processes

$$-\frac{dc_A}{dt} = \frac{dc_B}{dt} = kc_Ac_B. \qquad (6.39)$$

Let us solve these differential equations with some plausible initial conditions. Be $c_{A,o}$ the initial concentration of A and $c_{B,o}$ that of B. Let us introduce the decrease of the concentration of A during the course of reaction as a variable and denote it by x. The rate equation can then be written as

$$-\frac{d(c_{A,o} - x)}{dt} = \frac{d(c_{B,o} + x)}{dt} = k(c_{A,o} - x)(c_{B,o} + x). \qquad (6.40)$$

Performing the derivations at the left-hand sides (considering that both $c_{A,o}$ and $c_{B,o}$ are independent of time), we get a somewhat simpler form:

$$\frac{dx}{dt} = k(c_{A,o} - x)(c_{B,o} + x). \qquad (6.41)$$

The only difference between this differential equation and Eq. (3.55) in Sect. 3.1.3 is that, instead of $c_{B,o} - x$ there, we get $c_{B,o} + x$ here. Otherwise, the solution here goes the same way as there; we should solve the separated equation

$$\int \frac{1}{(c_{A,o} - x)(c_{B,o} + x)} dx = \int k \, dt. \qquad (6.42)$$

Let us resolve the rational algebraic fraction on the left side into the sum of partial fractions:

$$\int \frac{1}{(c_{A,o} - x)(c_{B,o} + x)} dx = \frac{1}{c_{A,o} + c_{B,o}}$$
$$\times \int \left(\frac{1}{(c_{A,o} - x)} + \frac{1}{(c_{B,o} + x)} \right) dx. \qquad (6.43)$$

Now we can readily integrate both sides to get the general solution

$$\frac{1}{(c_{A,o} + c_{B,o})} \ln \frac{c_{A,o}(c_{B,o} + x)}{c_{B,o}(c_{A,o} - x)} = kt + I. \qquad (6.44)$$

Making use of the initial conditions, rearranging and making a few substitutions we get the following explicit particular solution, similarly to that in Sect. 3.1:

$$c_B = \frac{c_{A,o} + c_{B,o}}{1 + \frac{c_{A,o}}{c_{B,o}} e^{-(c_{A,o} - c_{B,o})kt}}, \text{if } c_{A,o} > c_{B,o}. \qquad (6.45)$$

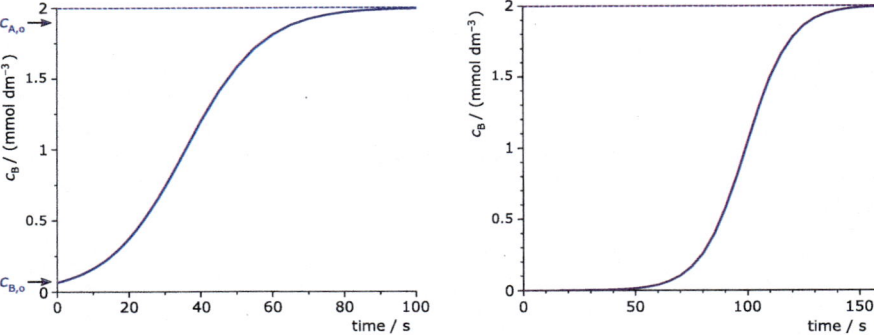

Fig. 6.6 Concentration of the component B (product) as a function of time during the course of the autocatalytic reaction A + B → 2 B. The rate coefficient k is 0.05 dm^3 mmol^{-1} s^{-1}, initial concentrations in the left panel are $c_{A,o} = 1.93$ mmol/dm^3 and $c_{B,o} = 0.07$ mmol/dm^3, also marked in the diagram. The horizontal dashed line is the limit of the concentration c_B ($c_{B,\infty} = c_{A,o} + c_{B,o}$) after the completion of the reaction. The 'sigmoid' shape (The word sigmoid is an allusion to the form of the Greek letter σ.) is characteristic of autocatalytic reactions but its proportions depend on the ratio of the concentrations of A and B. For example, in the right panel $c_{A,o} = 1.9999$ mmol/dm^3 while B is only present in a very low concentration: $c_{B,o} = 0.0001$ mmol/dm^3. Due to this very low catalyst concentration, the reaction starts extremely slowly. However, as soon as the concentration of B increases to 0.07 mmol/dm^3, further evolution of the curve is identical to that shown in the left panel

It is worth analysing the course of the reaction by following the time-dependent concentration of the product B (cf. Fig. 6.6). When the initial concentration of the autocatalyst B is small, the reaction begins quite slowly. It then accelerates more and more as the concentration of B increases, which continues until an inflexion point where the rate is maximal – since where the second derivative is zero, the first derivative has an extremum. After the inflexion point the increasing tendency of rate switches into a decreasing one, when the concentration of A becomes less than the actual concentration of B. The rate continues to decrease until the entire amount of A is transformed into B. As a result of the stoichiometry, the concentration of the product B after the completion of the reaction is equal to the sum of the initial concentrations $c_{A,o}$ and $c_{B,o}$.

The solution has some interesting properties. If the initial concentration of B is zero, the reaction would not even start. (This is a natural consequence of the fact that the reaction cannot proceed without the catalyst B.) This property is easily seen already in the rate equation (6.39): if $c_B = 0$, the rate is also zero. Similarly, solution (6.45) also gives zero as result at all time instants if $c_{B,o} = 0$. We can also see that solution (6.45) only holds if $c_{A,o} > c_{B,o}$; otherwise – if in the contrary, $c_{A,o} < c_{B,o}$ – we should switch the initial concentrations of A and B in the fraction, as well as in the exponent:

$$c_B = \frac{c_{A,o} + c_{B,o}}{1 + \frac{c_{B,o}}{c_{A,o}} e^{-(c_{B,o}-c_{A,o})kt}}, \text{if } c_{A,o} < c_{B,o} \quad (6.46)$$

This property of the solution is easily explained by the fact that the roles of A and B are symmetrical in rate equation (6.39). For this reason, we can *formally* interpret the overall reaction A + B → 2 B thinking that A is catalysing the reaction B → 2 B. However, as A is consumed, the definition of a catalyst strictly speaking does not hold concerning component A.

We can ask the question whether the condition that a pure substance A does not contain substance B *at all* was lifelike. Even if we allow for such purity of A, it could slowly be transformed in a small portion to B after interaction with, for example light, cosmic rays or heat. Be it a small portion only, the generated molecules of B would catalyse their further formation which process would be accelerated due to the autocatalytic effect. Similar cases can occur, among others in atmospheric chemistry or in physiological processes. As an example, we can see the right panel of Fig. 6.6, where the concentration of B is originally only 0.1 μmol/dm^3, less than 1/10000 part of the concentration of A.

Autocatalytic reactions can be much more complex than this simple example. For a composite reaction to be autocatalytic, it is sufficient that in one step of the mechanism an autocatalyst is formed which then catalyses other connected reaction steps.

6.3.2 Oscillating Reactions

Autocatalysis and autoinhibition – due to their positive and negative feedback – can lead to exotic chemical consequences. One of the most interesting and basic cases is chemical oscillation. In an *oscillating chemical reaction*, the concentration of at least one component – typically an intermediate – changes as a periodic function in time; that is, not changing monotonically but passing through subsequent minima and maxima. (At the same time, reactant concentration decreases and product concentration increases according to step-like functions.) As it will be shown, in order for a reaction to oscillate there should be nonlinear terms in the rate equation with respect to concentrations. For this reason, chemists usually call these and related composite reactions as *nonlinear chemical processes*.

Nonlinear processes occur in many different non-chemical systems as well. The first system of differential equations that shows close similarity to those describing oscillating reactions has been conceived for the description of oscillating animal populations. From the first researchers modelling oscillatory population dynamics, a simple model is called the Lotka-Volterra equations.[11] Volterra has modelled the periodic fluctuation of some Adriatic fish populations; that of the predator fishes and their prey fishes oscillated periodically but in different phases, sometimes increasing, other times decreasing.

Marine fish population has a multitude of species that were divided into two classes by Volterra: predator fishes and prey fishes. However, it is easier to demonstrate this model in an inland version with only two animal species. Accordingly, let us suppose that there are only two vertebrate species in a sufficiently great island: foxes (predators) and rabbits (preys). The island is great enough that there is always plenty of grass for rabbits and plenty of places suitable for nesting (digging holes) for both rabbits and foxes. For the sake of simplicity, let us also suppose that reproduction of both rabbits and foxes means that from one of them we get two of them (a quadratic nonlinearity); and that rabbits cannot survive until natural death as they will be consumed by foxes before. Another simplifying condition is that the consumption of one rabbit by a fox is sufficient for producing one offspring. Finally, we suppose that the rate of capturing a prey, reproduction and death are proportional to the size of the population. Denoting rabbits by R, foxes by F, and imitating stoichiometric equations we can write the corresponding mechanism:

$$R \xrightarrow{k_1} 2R \quad \text{reproduction of rabbits} \quad (6.47)$$

$$R + F \xrightarrow{k_2} 2F \quad \text{reproduction of foxes} \quad (6.48)$$

$$F \xrightarrow{k_3} \varnothing \quad \text{death of foxes} \quad (6.49)$$

Rate equations of the above mechanism are the following:

$$\frac{d[R]}{dt} = 2k_1[R] - k_2[R][F] \quad (6.50)$$

$$\frac{d[F]}{dt} = 2k_2[R][F] - k_3[F] \quad (6.51)$$

Nonlinear behaviour is the consequence of the product [R][F] found in both rate equations. The above system of differential equations does not have an analytical solution; the solution can only be obtained by numerical integration after specifying the initial conditions of the population densities [R] and [F]. Figure 6.7 shows such a

---

[11] Alfred James Lotka (1880–1949) was an American biophysicist and statistician who was born in Lemberg – then Austria-Hungary, nowadays Lviv, Ukraine – and whose parents were citizens of the United States. He began his studies in Europe but finished them in the United States and lived there ever after. He wrote in 1925 about predator–prey relations in his book on population dynamics entitled *Elements of Physical Biology*. Vito Volterra (1860–1940) was an Italian mathematician and physicist who was born in Ancona (by that time belonging to the Papal States). Independently of Lotka, he published his results in 1926 interpreting data collected by his fish biologist son-in-law concerning large periodic oscillations in the fish population of the Adriatic sea.

6.3 Autocatalysis, Autoinhibition and Nonlinear Chemical Processes

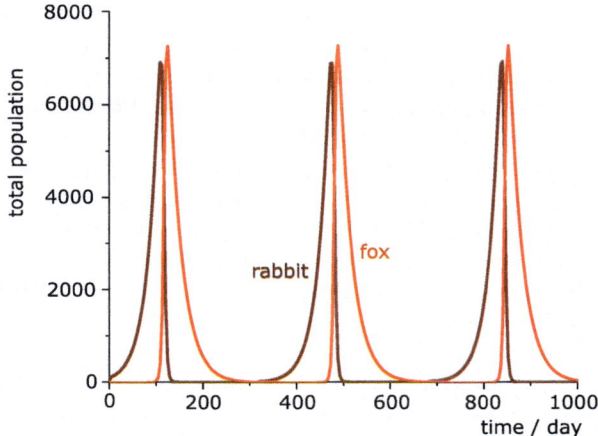

Fig. 6.7 Solution of the Lotka-Volterra rate equations (6.50)–(6.51). Starting population densities are 5 km^{-2} of rabbits and 0.1 km^{-2} of foxes. Rate coefficients: $k_1 = 0.04$ day^{-1}, $k_2 = 0.001$ day^1 and $k_3 = 0.035$ day^{-1}. Rabbit and fox population of the island of 10 km^2 surface oscillate roughly with a period of one year. Note the delay of increase and decrease of the fox population with respect to that of the rabbits. The shape of increase of both populations is similar to that of the product in the simple autocatalytic reaction discussed before

solution with the parameters given in the figure caption. (It is worth noting that population dynamics is more complex than this; for the sake of transparency and simplicity we have used several conditions that are not necessarily relevant to the behaviour and life cycle of rabbits and foxes.)

The origin of the oscillatory behaviour is the fact that the small fox population cannot multiply fast enough until rabbits are also scarce. This gives the possibility for rabbits to multiply at a considerable rate. However, as there will be more and more rabbits, foxes multiply also faster and faster; after a while, they will be numerous enough that they can consume rabbits faster than rabbits could multiply, which leads to a fast decrease of the rabbit population. Decreasing rabbit population leads – with a certain delay – to large death rate of the starving fox population. When there are only a few rabbits and a few foxes again, the cycle can rebegin. Both multiplication processes R → 2 R and R + F → 2 F are equivalent to autocatalytic processes in chemical reactions, manifested in a slow initial rate of increase in population which accelerates continuously. Though there is no chemical reaction which would be equivalent to the Lotka-Volterra mechanism (6.50)–(6.51), this model is simple and transparent to easily follow the effect of feedback in case of nonlinear rate equations.

It is important to note that the population dynamic model discussed does not represent a closed system; grass necessary for rabbits to live and multiply grows continuously in a sufficient quantity, and nesting places are also always available. If we imagine an equivalent chemical system, there should be a continuous external supply of reactants and removal of products. Such systems are chemically open; oscillation with no decrease in the amplitude can only occur in open systems. (A frequent and popular experimental realisation of open chemical oscillations is the so-called *continuously-fed stirred tank reactor* – abbreviated as CSTR – where reactants are continuously loaded and part of the reaction mixture drained.) Chemical oscillation in a closed system is continuously damped due to dissipation, until it stops completely.

Oscillating composite reactions are not very rare; poorly controlled older combustion engines with lower quality fuel were often 'knocking' at lower revolution speed due to spontaneous oscillatory burning of the air/fuel mixture in the cylinder. In living organisms, there are a large number of periodic phenomena (*e.g.* heartbeat, breathing); many of them regulated (at least in part) by oscillating chemical processes. For example, in physiological processes, the value of pH often plays an important role. One of the simplest oscillating chemical systems is the oscillation of the concentration of the hydrogen ions in the following mechanism, first described by Rábai and Beck and named as *alternator*:

$$A + B \to Y \qquad (6.52)$$

$$A + B + X \to P_1 \qquad (6.53)$$

$$A + Y + X \to 2X + P_2 \qquad (6.54)$$

In this mechanism, A refers to IO_3^-, B to $S_2O_3^{2-}$, X to H^+, Y to HSO_3^-, P_1 to $S_4O_6^{2-}$ and P_2 to SO_4^{2-} in their dissociated or undissociated form. We can see that the mechanism is quite similar to the Lotka-Volterra model; however, due to its somewhat greater complexity, this can be realised in a chemical system. The reason for oscillation is the alternation of the autocatalytic formation of X in step (6.54) and its pseudo-first order disappearance in step (6.53), coupled to the formation of Y in (6.52). (Detailed stoichiometry of the reactions involved is quite complicated; the structure of the processes leading to oscillation – the so-called *skeleton model* – is easier to see in this simplified way of writing.)

Nonlinear chemical processes do not only result in temporal oscillation; they can also result in periodic spatial pattern formation. As examples, patterns of solid exoskeletons or those of coloured butterfly wings are thought to develop based on nonlinear chemical phenomena.

6.3 Autocatalysis, Autoinhibition and Nonlinear Chemical Processes

Problems
1. The rate-determining step in ammonia synthesis over Ru catalysts is believed to be the dissociative adsorption of N_2 because the reaction is first order in N_2, no H_2/D_2 isotopic effect is observed, and the active binding sites on ruthenium are occupied overwhelmingly by hydrogen atoms. (see Bécue et al., in Further Reading). According to these conditions, the kinetics of ammonia synthesis can be described by a reaction mechanism with dissociative adsorption of N_2 as the rate determining step, and adsorbed hydrogen atoms as the most abundant reactive intermediates on the surface. Dissociative adsorption of H_2 can be considered to be in equilibrium, and the formation of ammonia on the surface together with its desorption to the gas phase (several reaction steps) can also be considered to be in equilibrium – as they both follow the rate determining step:

$$N_2 + 2\,S \xrightarrow{k_1} 2\,NS \quad \text{rate determining step}$$

$$H_2 + 2S \rightleftharpoons 2HS \quad \text{equilibrium constant}: K_2$$

$$HS + 3HS \rightleftharpoons NH_3 + 4S \quad \text{equilibrium constant}: K_3$$

where S is the active site binding either an H atom, an N atom or other intermediates of the reaction leading to ammonia. For some Ru catalyst preparations, it was found for the forward rate of ammonia synthesis on a uniform surface that the order in N_2 is unity, in H_2 it is -1, and zero in ammonia.

Show that this mechanism and the given conditions can rationalise the experimental kinetics found.

Solution: Let us use similar notation as with the derivation of the kinetics of reaction (6.6): c is the overall concentration of active centres on the surface (occupied or not), $c\,(1-\theta_H)$ is the concentration of non-occupied (free) active centres on the surface, c_{N_2} is the gas phase concentration of N_2 and c_{H_2} is that of H_2. The concentration of HS can be written as $c\,\theta_H$, where θ_H is the fraction of active centres on the surface covered by adsorbed H atoms. (Other surface species have negligible coverage on the surface.)

As the first step determines the reaction rate, we can write

$$r = -\frac{dc_{N_2}}{dt} = k_1 c_{N_2} c^2 (1-\theta_H)^2.$$

The equilibrium constant K_2 can be written as $K_2 = \frac{c^2 \theta_H^2}{c^2(1-\theta_H)^2 c_{N_2}}$, from which we can express $c^2(1-\theta_H)^2 = \frac{c^2 \theta_H^2}{K_2 c_{N_2}}$.

Substituting this into the expression of the reaction rate r we get $r = \frac{k_1 c^2 \theta_H^2}{K_2} \frac{c_{N_2}}{c_{H_2}}$, where the first fraction could be changed with changing θ_H^2, but the fact that H atoms occupy most of the active sites, and their concentration would not change much with the change of c_{H_2}, we can simplify the equation by setting $\theta_H \cong 1$ to get $r \cong \frac{k_1 c^2}{K_2} \frac{c_{N_2}}{c_{H_2}}$, which has the property of the experimental findings: it is first order in N_2, of order -1 in H_2, and zero order in NH_3.

2. The enzyme invertase hydrolyses sucrose (sugar cane, a disaccharide) to glucose and fructose. The reaction can be followed by the detection of optical rotation, as the specific rotation of the dextrorotatory sucrose of 66.5 deg cm^3 g^{-1} dm^{-1} changes to the sum of the specific rotations of the two products of -39.7 deg cm^3 g^{-1} dm^{-1} during the course of reaction.

In a series of experiments, time-dependent optical rotation θ was followed in a 1 dm long optical path polarimeter at 20 °C, using 589 nm sodium D-light, and the initial rate of this change has been determined at the same enzyme concentration but different sucrose concentrations [S]. The molar concentration of the enzyme was at least an order of magnitude lower than that of the substrate. The following data were obtained:

[S]; mol dm^{-3}	dθ/dt; deg min^{-1}
0.0052	−0.00919
0.0104	−0.01485
0.0208	−0.0215
0.0416	−0.0276
0.0833	−0.0322
0.167	−0.0352
0.333	−0.03685

Calculate the parameters r_∞ and K_M for this enzyme-catalysed reaction.

Solution: Let us consider the Michaelis-Menten expression (6.34) for the rate of product formation $\frac{d[P]}{dt} \cong \frac{r_\infty [S]}{K_M + [S]}$. We need to determine the rate of product formation as mol dm^3 min^{-1}; thus we have to convert values reported in deg min^{-1}. To do so, we can use the relation between the angle of rotation of the plane-polarised light and the concentration of optically active species. In this case, this relation reads as $\theta = [\alpha]_{D;S}^{20}[S] \ell + [\alpha]_{D;P}^{20}[P] \ell$, where $[\alpha]_{D;S}^{20} = 66.5$ deg cm^3 g^{-1} dm^{-1} is the specific rotation of the substrate sucrose, $[\alpha]_{D;P}^{20} = -39.7$ deg cm^3 g^{-1} dm^{-1} that of the products, and ℓ is the optical path length of the polarimeter cuvette the plane-polarised light passes through.

Let us make use of the stoichiometry of the reaction: be $x = [P]$ the molar concentration of the product (not being present at the start of the reaction); then the substrate concentration can be written as $[S] = [S]_o - x$, where $[S]_o$ is the initial

6.3 Autocatalysis, Autoinhibition and Nonlinear Chemical Processes

concentration of S. Using this notation, we can rewrite the equation for the actual optical rotation the following way:

$$\theta = [\alpha]_{D;S}^{20}([S]_o - x)\ell + [\alpha]_{D;P}^{20} x \ell.$$

Solving this equation for x we get

$$x = \frac{[\alpha]_{D;S}^{20}[S]_o}{[\alpha]_{D;S}^{20} - [\alpha]_{D;P}^{20}} - \frac{1}{\left([\alpha]_{D;S}^{20} - [\alpha]_{D;P}^{20}\right)\ell}\theta.$$

We still need to harmonise units of quantities; specific rotations need to be converted in units in agreement with molar concentrations: deg dm^3 mol^{-1} dm^{-1}. The conversion factor with respect to dm^3 as volume and mole as quantity is 1000/M, where M is the molar mass in gram/mol units. Thus, we should use 194.27 deg dm^3 mol^{-1} dm^{-1} for $[\alpha]_{D;S}^{20}$ and -110.18 deg dm^3 mol^{-1} dm^{-1} for $[\alpha]_{P;S}^{20}$. The rate of the product formation can be given as dx/dt, thus the searched-for conversion of the reaction rate can be written as

$$\frac{d[P]}{dt} = \frac{dx}{dt} = -\frac{1}{\left([\alpha]_{D;S}^{20} - [\alpha]_{D;P}^{20}\right)\ell}\frac{d\theta}{dt}$$

$$= -\frac{1}{(194.27 + 110.18)\text{deg dm}^3\text{mol}^{-1}\text{dm}^{-1} \cdot 1\text{dm}}\frac{d\theta}{dt}.$$

Next we have to convert the rate of change of the optical rotation to the reaction rate and use the Michaelis-Menten equation cited above to estimate its parameters from the experimental data. After the conversion described above, we get the following table:

[S]; mol dm^{-3}	$d\theta/dt$; deg min^{-1}	$d[P]/dt$; mol dm^3 min^{-1}
0.0052	−0.00919	3.019 × 10^{-5}
0.0104	−0.01485	4.878 × 10^{-5}
0.0208	−0.0215	7.062 × 10^{-5}
0.0416	−0.0276	9.066 × 10^{-5}
0.0833	−0.0322	1.058 × 10^{-4}
0.167	−0.0352	1.156 × 10^{-4}
0.333	−0.00919	1.210 × 10^{-4}

Least-squares estimation provides the following 95% confidence interval for the parameters: $r_\infty = 1.271 \times 10^{-4} \pm 0.002 \times 10^{-4}$ s^{-1} and $K_M = 0.0167 \pm 0.0001$ mol dm^3. The high precision is due to the fact that polarimetric measurements are rather precise and the reaction is quite slow compared to the time necessary for mixing reactants and reading the rotation angle. Accordingly, the fit of the model function $\frac{d[P]}{dt} = \frac{r_\infty[S]}{K_M+[S]}$ is also very good. The good fit also

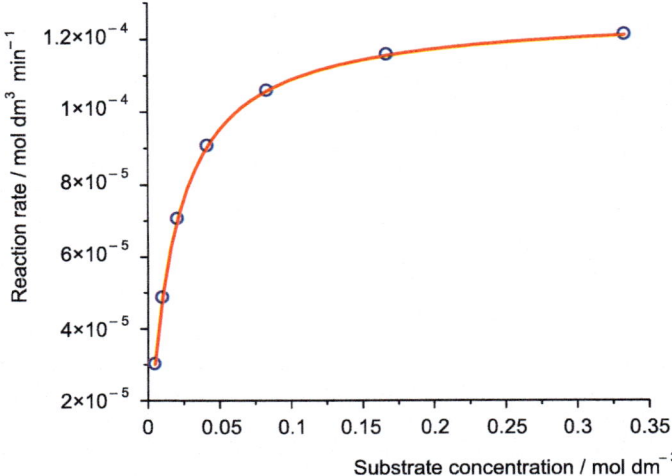

Fig. 6.8 Quality of the fit of the Michaelis-Menten mechanism with the estimated parameters. Blue circles show measured data, continuous red curve the fitted reaction rate with the estimated parameters

indicates that the reaction follows the Michaelis-Menten mechanism well. Figure 6.8 shows the actual fit (red line) to the experimental points (blue circles):

3. Show that the ratio of substrate concentrations [S] at two different initial rates pr_∞ and qr_∞ of an enzyme-catalysed reaction according to the Michaelis-Menten mechanism is independent of the parameters of r_∞ and K_M. Calculate this ratio for $p = 0.9$ and $q = 0.3$.

Solution: Let us write the reaction rate r as a function of r_∞ and K_M according to Eq. (6.34):

$$\frac{d[P]}{dt} = r \cong \frac{r_\infty [S]}{K_M + [S]} = \frac{r_\infty}{\frac{K_M}{[S]} + 1}$$

We can solve this equation for [S] and get $[S] = \frac{r}{r_\infty - r} K_M$. Now we can write the equation for the ratio of the two substrate concentrations:

$$\frac{[S]_p}{[S]_q} = \frac{\frac{pr_\infty}{r_\infty - pr_\infty}}{\frac{qr_\infty}{r_\infty - qr_\infty}} = \frac{\frac{p}{1-p}}{\frac{q}{1-q}} = \frac{p(1-q)}{q(1-p)}.$$

As we can see, the result is indeed independent of both of r_∞ and K_M. We can note that this offers a simple way to check if the mechanism of the enzyme-catalysed reaction follows the (simple) Michaelis-Menten mechanism, or it is more complicated.

Using the above result, it is straightforward to calculate the ratio of the substrate concentrations at the reaction rates of 0.9 r_∞ and 0.3 r_∞; it is 21; thus, the substrate concentration at which the initial rate of the reaction is 0.9 r_∞ is 21 times greater than the substrate concentration at which the initial rate is 0.3 r_∞.

Further Reading

1. Pilling MJ, Seakins PW (1995) Reaction kinetics. Oxford University Press, Oxford
2. de Paula J, Atkins PW (2014) Physical chemistry. 10th edn. Oxford University Press, Oxford
3. Silbey LJ, Alberty RA, Moungi GB (2004) Physical chemistry. 4th edn. Wiley, New York
4. Steinfeld JI, Francisco JS, Hase WL (1998) Chemical kinetics and dynamics. 2nd edn. Prentice Hall, Englewood Cliffs
5. Ross JRH (2012) Heterogeneous catalysis: fundamentals and applications. Elsevier, Amsterdam
6. Lehninger AL, Cox MM, Nelson DL (2017) Lehninger principles of biochemistry. 7th edn. Freeman, New York
7. Bagshaw CR (2017) Biomolecular kinetics: a step-by-step guide. CRC Press, Boca Raton
8. Mikhailov AS, Ertl G (2017) Chemical complexity: self-organization processes in molecular systems. Springer, Cham
9. Rábai G, Beck MT (1988) High-amplitude hydrogen ion concentration oscillation in the iodate-thiosulfate-sulfite system under closed conditions. J Phys Chem 92:4831–4835
10. Press WH, Teukolsky SA, Vetterling WT, Flannery BP (2007) Numerical recipes: the art of scientific computing. 3rd edn. Cambridge University Press, New York
11. Bevington PR, Robinson DK (2002) Data reduction and error analysis for the physical sciences. 3rd edn. McGraw-Hill, Boston
12. Bécue T, Davis RJ, Garces JM (1998) Effect of cationic promoters on the kinetics of ammonia synthesis catalyzed by ruthenium supported on zeolite X. J Catal 179:129–137

Chapter 7
Experimental Methods in Reaction Kinetics

Reaction kinetics is concerned with the temporal evolution of chemical reactions. Most important goal of kinetic experiments is to explore and identify a correct mechanism of reactions. For this, it is necessary to identify all components involved in the reaction and follow the temporal evolution of their concentration. Composite reactions consist of elementary reaction steps; thus it is of crucial importance to determine their rate coefficients as a function of temperature, pressure and sometimes other conditions (e.g. different solvents, or inert gas content). For this reason, the typical task in experimental kinetics is to determine the time-dependent concentration of components under controlled conditions. To fulfil this task, we need devices where reactions can be run under known conditions (these devices are called *reactors*) and instruments to measure signals related to concentration (these are called *detectors*). In a kinetic experiment, mixing of the reactants is very important; possibly within as short a time while the reaction is not proceeding in a detectable extent. In summary, an experimental kinetic device should be a reactor where mixing can be done quickly enough, and it should have detectors that can provide signals within sufficiently short time to determine the concentration of reacting components precisely enough. In addition, we should be able to measure time with necessary precision during the course of reaction.

One of the most important characteristics of the reactor and the detectors is their *time resolution*. This means that mixing of reagents as well as measuring actual concentrations should happen within short enough time in order that the reaction would not advance in a detectable extent. For this reason, let us consider the characteristic time of different kinds of chemical reactions. In Fig. 7.1, we can see their timescales. The longest time available on Earth is that of the age of the planet. This timescale is characteristic of reactions taking place in the Earth's crust within its solid components (minerals). We can have indirect information concerning these reactions, based on present measurements. The same is valid for much shorter reactions but still at a timescale of more than a few months; their thorough real-time observation is not easy due to the usual work schedule of researchers. The timescale of easy study of reactions spans from a few days to a few seconds.

© The Author(s), under exclusive license to Springer Nature Switzerland AG 2021
E. Keszei, *Reaction Kinetics*, https://doi.org/10.1007/978-3-030-68574-4_7

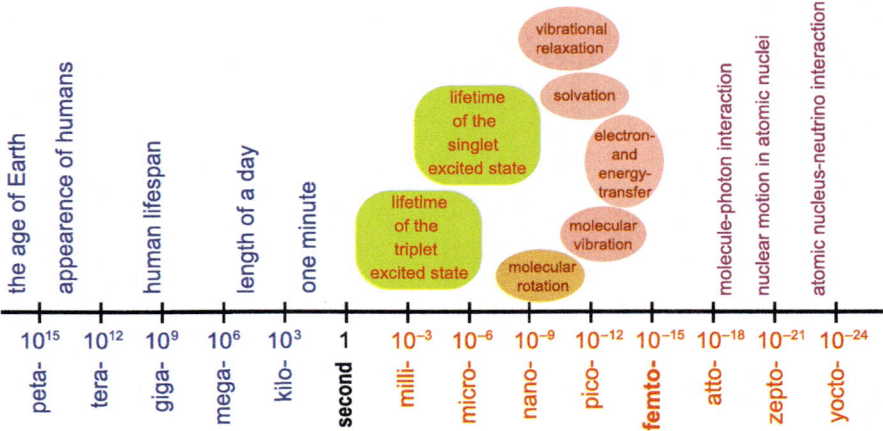

Fig. 7.1 Timescale of chemical reactions. Only reactions shorter than the age of the Earth can be studied on Earth – most of them only with indirect methods. Between 1850 and 1930, the available timescale to follow reactions was between a few seconds and a few days. Molecular events of reactions happen within the timescale from milliseconds to femtoseconds; their study is conveniently carried out using short laser pulses. At still shorter times, only processes concerning molecular and nuclear physics can happen. An interesting feature of the chemically relevant timescale is that it spans over 15 orders of magnitude both longer and shorter than 1 s

Reactions that go to completion in less than 1 s need special methods and devices, concerning both mixing their reactants and measuring concentrations of reacting components. Measuring clock time shorter than a nanosecond is no more feasible, even if we use the fastest clocks; thus, we need some alternative methods to determine reaction time at the picosecond or even shorter timescale.

As explained in Sect. 2.2 when discussing transition state theory, molecular events of chemical reactions cannot happen quicker than the characteristic time of a vibrational motion, of which the shortest limit is roughly 10^{-14} s; thus, shorter times than 1 fs (1 fs = 10^{-15} s) are not relevant concerning reactions. Though the interaction of molecules with photons happens within attoseconds (1 as = 10^{-18} s), only the electronic structure of the molecule is able to change within this time; to change the position of nuclei – which is the definition of a proper chemical reaction – at least a few femtoseconds are needed. Nucleon motion within atomic nuclei and nucleus–neutrino interactions do not induce immediate chemical changes; chemical reactions occur at a much slower pace following transformation of atomic nuclei.

In the following sections, an overview of the possibility to follow concentration as a function of time, and the description of the problem to start reactions is given first. Different solutions for measuring time are discussed in detail when describing reactors and detectors.

7.1 Methods to Determine Concentration

Concentration can be determined by measuring some quantity (called the *detector signal*) which is a well-known function of concentration. In the simplest case, this quantity is directly proportional to the concentration. As this is not always the case, a common practice is to trace a calibration curve at sufficiently close known concentration values and determine an interpolating function. A linear function of concentration of the detector signal is especially useful in case of first-order (or pseudo-first-order) reactions; in this case, the rate coefficient can be determined without the knowledge of the initial concentration and its absolute value, simply based on the rate of change of concentration (see Sect. 3.1.2). Rearranging Equation (3.34), we get $c(t)/c_o = e^{-kt}$; from this expression we can see that it is possible to determine the rate coefficient k from the ratio $c(t)/c_o$ as a function of time, and time can be measured from any instant during the course of reaction. (If the detector signal is directly proportional to the concentration, the ratio of the measured signals at time t and time zero is equal to $c(t)/c_o$.) If parameters of the reaction (e.g., rate coefficients or initial concentrations) are dependent also on the absolute value of concentration, it is necessary to calculate it from the detector signal.

In kinetic experiments, optical detection methods are very often used. These provide information on concentration by some optical properties changed in interaction with the reaction mixture. Interaction of light with matter depending on concentration can be *refraction*; *absorption* or *emission* of light; *rotation* (chiral molecules can rotate the plane of polarisation of plane-polarised light when passing through the sample) and *circular dichroism* (chiral molecules can absorb right and left circularly polarised light differently). *Reflection* of light can be used to measure surface concentration.

If the concentration in a solution is sufficiently low, the *refractive index* is proportional to concentration. To measure it in situ, the reaction mixture should be an optical element of the *refractometer*; for this reason, its use is limited in kinetics. In case of slow reactions (reaction times in the order of an hour) a small sample can be withdrawn from the reaction mixture and put into the refractometer. A severe limitation is that refraction can only measure the total concentration of mixtures.

Optical rotation can be measured using a *polarimeter*. It is not complicated to use in kinetic measurements; the reactor can be placed between the *polariser* and the *analyser* of the polarimeter. This method can only be used to determine the concentration of optically active components. The angle of rotation of the plane-polarised light is usually proportional to the concentration of the optically active component.[1]

---

[1] The very first quantitative kinetic measurement was made using a polarimeter. Ludwig Ferdinand Wilhelmy (1812–1864) German physical chemist determined the rate coefficient of acid-catalysed hydrolysis of saccharose (common table sugar). Saccharose is dextrorotatory (rotates the plane of polarised light right), while the products of hydrolysis – glucose and fructose – laevorotatory (rotate it left). (This is the reason to call this process *sugar inversion*.) The paper reporting results has been published in the journal *Annalen der Physik und Chemie*, volume 81, pp. 413–433 in 1850.

A *CD spectrometer* measuring circular dichroism is only rarely used in kinetic experiments.

One of the simplest and most widely used methods to determine concentration is based on the absorption of light, measured by an *absorption spectrophotometer*. The principle of this instrument is to measure the intensity I_o of the incident light into the sample and the intensity I of the transmitted light through it, at a single wavelength λ selected by the monochromator of the spectrophotometer. From these two intensities we can calculate the concentration (if it is not too high) based on the Beer–Lambert law[2]:

$$A^\lambda = -\log_{10}\left(\frac{I}{I_0}\right) = \varepsilon^\lambda c \ell. \qquad (7.1)$$

In the equation, A^λ is the *absorbance* at the wavelength λ, ℓ is the *optical path* in the cuvette and ε^λ is the *decadic molar absorptivity* also at wavelength λ. In many cases, there are more than one species in the reaction mixture that absorb light at the wavelength of the measurement. In such cases, the measured absorbance is the sum of the contributions of all the n absorbing components:

$$A^\lambda = \sum_{i=1}^{n} \varepsilon_i^\lambda c_i \ell. \qquad (7.2)$$

As the absorption spectrum of different components changes differently as a function of wavelength, to resolve particular concentrations, absorbance should be measured at least at n different wavelengths at the same time instant. If the contribution to the absorbance (molar absorptivities) of the components is different (and known) at each wavelength, the concentration of the n species can be calculated. (Unknown molar absorptivities can be estimated from measured data.)

Besides this additive property, absorbance detection has another disadvantage. In cases when a small change in absorbance should be measured in an otherwise highly absorbing mixture due to small concentration changes, the uncertainty (noise) of the large signal can be comparable to the small change. A convenient alternative is to measure only the change, against a zero signal. This is possible if we measure *emitted light* instead of the transmitted one. To get emitted light from a given component, the mixture should be irradiated by (typically ultraviolet) light that *selectively excites* this component, and – while getting de-excited – it emits light at a characteristic (usually lower than the exciting) wavelength. De-excitation typically

---

[2]French mathematician and physicist Pierre Bouguer (1698–1758) was the first (prior to 1729) to recognise the exponential law of light intensity attenuation when it passes across atmospheric air. Johann Heinrich Lambert (1728–1777) Swiss mathematician and physicist cited this result and applied the law in a publication in 1760. It was August Beer (1825–1863) German physicist and chemist who related the exponential attenuation to concentration in solutions. The usual name in English literature is Beer–Lambert law, but it is also called Beer's law or Lambert–Beer law. Less frequently, the name Beer–Lambert–Bouguer law is also used.

7.1 Methods to Determine Concentration

results in a state of the same multiplicity than that of the excited state and the emission is called *fluorescence*. Excited state lifetime of molecules prior to emission (the *fluorescence lifetime*) is quite short: it is in the nanosecond to microsecond time range. Thus, the fluorescence detection method – in addition to its sensitivity and selectivity – is a suitable method when high temporal resolution is needed. Fluorescence measurements can be done using *fluorimeters* which have similar construction to absorption spectrophotometers but typically have much higher light sensitivity. As excitation light source, a simple discharge tube with continuous emission of UV light can be used, but lasers are more convenient for this purpose. In the twenty-first century, high intensity tunable pulsed lasers are easily available, emitting nearly monochromatic, very short pulses; their pulse width may be even as short as a few femtoseconds. In case of de-excitation into a state of different multiplicity, the emission is called *phosphorescence*. Phosphorescence lifetime is typically longer than that of fluorescence (due to the forbidden transition between different multiplicity states): from microseconds even up to a few seconds. Detection of phosphorescence is also carried out by fluorimeters.

A special case is the so-called *resonance fluorescence* which occurs if the excited molecule is very unstable (the fluorescence lifetime is very short) and does not have much low-lying energy levels. These are typically atoms, or radicals consisting of few atoms only. To excite resonance fluorescent states, the first light sources were discharge tubes filled with the same material as the component whose concentration should be determined. As the process in the discharge tube is the same (following excitation, return to the same state as before excitation), the energy of photons emitted is exactly the same as that of those which can excite these molecules in the reaction mixture. The name originates in the fact that the emitted photons from the discharge tube have exactly the same energy as the absorbed ones in the reaction mixture; thus, the condition for resonance is fulfilled.

Whether we measure absorbance, fluorescence or phosphorescence, *photodetectors* are necessary to measure light intensity. These photodetectors can be *photomultiplier tubes*, or some semiconductor devices (e.g. *photoresistors* or *photodiodes*). Photomultiplier tubes (abbreviated as PMT) are devices used since the early years of photodetection, which have a great sensitivity; some of them able to detect one single photon. Photoelectrons are generated by photons hitting the surface of photocathodes, which in turn are multiplied by a cascade of so-called *dynodes* with high voltage between them; thus, after the impact of one photon on the photocathode, as much as 10^8 electrons can be detected. However, semiconductor-based photodetectors developed in the twenty-first century also approximate the one-photon sensitivity; thus, PMTs which need high voltage are less and less used recently to follow reactions.

Many concentration detection methods are based on other than optical properties. To measure ion concentration, *ion-selective electrodes* are appropriate to use. Concentration of neutral species can also be measured using *membrane electrode assemblies* based on redox reactions. To determine the overall concentration of all ionic species, a *conductivity meter* gives a convenient possibility. If there are gas phase ions in the reaction mixture, their concentration can also be measured using high enough voltage between two electrodes: the resulting ion current is proportional to the concentration of gaseous ionic species.

A convenient way to measure the concentration of radicals is to detect them based on the magnetic properties of their unpaired electrons. The most popular method is *electron spin resonance* (*ESR*) or *electron paramagnetic resonance* (*EPR*) based on the splitting of energy levels of the spin of unpaired electrons in a magnetic field (the *Zeeman effect*). Excitation is done by electromagnetic radiation around 9–10 GHz frequency, which is the energy gap between the two spin states ('up' and 'down'). Due to the Boltzmann energy distribution, the lower state is more populated, thus it absorbs the radiation and gets excited to the upper state. The absorption is proportional to the concentration of the radical. The frequency of the exciting pulse limits the temporal resolution in a few tenths of nanoseconds.

Nuclear magnetic resonance (*NMR*) spectroscopy operates on the same principle as EPR; the main difference is that the spin angular momentum of nuclei is roughly 1000 times smaller than that of the electron, which makes the splitting of states and the frequency of the exciting radiation accordingly smaller, in the tenths to hundreds of MHz region. However, with recent super-strong-field magnets, a few GHz splitting can also be achieved. NMR spectroscopy is eminently suitable to measure the concentration of organic substances, as protons have an unpaired nuclear spin. (In addition, isotopes ^{13}C and ^{15}N are also NMR active.) A great advantage is that the NMR signal offers the possibility of identifying individual components, and – unlike absorption spectroscopy where components have different molar absorptivities – each proton has the same molar absorptivity of the exciting radiation. Response time of the NMR excitation is also limited by the excitation frequency; thus, its time resolution is typically not shorter than a few ns.

To follow the time-dependent concentration of reactive species, *mass spectrometers* offer one of the best opportunities. Apart from the fact that they are able to determine the concentration of gas phase ions entering the ion source with time resolution of microseconds, unknown components can easily be identified by measuring their molar mass. Another convenient possibility is that if collision-free conditions are needed following the reaction (avoiding the change of state of the product molecules via collisions), the reactor can be placed directly into the vacuum system of the mass spectrometer. Mass spectrometers can select charged particles by their mass over charge ratio, thus they are ideal devices to study ion–molecule reactions. There are also large instruments where molecular beams are crossed in the reaction zone and the emerging products can be measured as a function of scattering angle, using movable mass spectrometers. The mass spectrometer is in the same vacuum system as the reaction zone, and it is fixed on a *goniometer*[3] to easily control directions of detection.

---

[3]The literal meaning of goniometer is an instrument to measure angle; it is coined from two ancient Greek nouns: γωνία (angle) and μέτρον (measure). In science, a controlling device which allows an object to be rotated to a precise angular position is also called a goniometer.

7.2 Initiating Reactions at Different Timescales

In case of a sufficiently long reaction (complete within several minutes or hours) we have enough time to fill into the reactor the prepared reactants and mix them to get a homogeneous phase. Mixing can be done either manually or with a common magnetic stirrer. (The stirrer should be carefully cleaned in both cases to avoid any catalytic or inhibitory interaction with a subsequent reaction mixture.)

If the reaction time is much shorter (a few minutes or seconds only), mixing should be very fast; completed well within a second. The best practice is to inject reactants with high pressure through some mixing device that can make a homogeneous mixture within a fraction of a second; thus mixing time can be neglected compared to reaction time. At this point, we can ask what is the temporal limit of mixing gaseous or liquid components (fluid phases). For this, we should know the time limit of making molecules to move in *one direction*.

In fluid phases, there exists a natural unidirectional motion: the propagation of sound. Sound is a longitudinal acoustic wave; thus, periodical decrease and increase in density of the medium propagates with the unidirectional velocity component of molecules. (This is the reason that the velocity of sound in a gas largely depends on temperature but only slightly on pressure.) We can say that the velocity of sound is closely related to the characteristic velocity of translational molecular motion. As a consequence, molecules of a fluid medium can readily be moved with a velocity less or equal to the velocity of sound; in case of a greater velocity, energy is needed not only to force unidirectional motion (instead of a random motion) but also to increase the thermal velocity of the molecules. (It is a similar phenomenon as when supersonic airplanes surpass the velocity of sound.) This can only be done with much higher energy input than it is possible using mixing devices. For this reason, mixing cannot be faster than the velocity of sound.

In dry ambient air of about 20 °C (at pressures not much different from 1 bar) the velocity of sound is 343 m/s (1235 km/h); in pure water under the same conditions it is 1440 m/s (5186 km/h). A mixing device to achieve fast mixing can be as small as 2 mm; thus, in ambient gas it means *cca.* 6 µs, while in water *cca.* 1.5 µs lower limit for mixing time. Mixing cannot be faster than this. Accordingly, reactions taking place in less than 10 µs cannot be launched by mixing.

Reactions shorter than a microsecond can only be initiated by indirect methods. This means that by mixing reagents, the reaction should not start. A simple way to do this is to mix one of the reactants in a non-reactive form called *precursor*.[4] The reaction would not start after mixing, only after a (very fast and short) interaction which would dissociate or excite the precursor resulting in the genuine reactant. A convenient interaction is an incident laser pulse; its temporal width, intensity and wavelength (the energy of photons) are easily controlled to have a desired value needed in the experiment. Other radiations (e.g. X-rays, electron beam or

---

[4]The literal meaning of the Latin word is 'forerunner'. It is a participle consisting of the preposition *prae* (before, in front of) and the verb *currere* (run, rush).

electromagnetic radiation of longer than visible light wavelength) can also be used, but laser pulses are very easy to control, thus they slowly superseded other pulsed energy sources. Using lasers, we can have access to pulses of duration from milliseconds to femtoseconds; thus, it is possible to initiate as fast reactions as we want.

7.3 Experimental Techniques and Devices

This section describes a variety of reactors and experiments which enable to study reactions at different timescales by measuring time-dependent concentrations. Some reactors contain a homogeneous mixture which remains in them until the end of the reaction. These are called *batch reactors*. Some other reactors accommodate a flowing reaction mixture (usually at constant flow rate) – these are called *flow reactors*. There are also reactors in which the reaction mixture is not homogeneously distributed; instead, a molecular beam (or two of them) is injected into the very low-pressure container under vacuum; or the reaction zone is along a track of high-energy radiation.

Details of initiation of reaction, detectors measuring time-dependent concentrations, available time resolution and the method of measuring reaction time are also discussed. Reaction rates – as well as rate coefficients – are temperature dependent; therefore, constant temperature experiments are appropriate to conduct using thermostats. However, we shall not discuss thermostats in connection with reactors.

7.3.1 Classical Techniques

Classical techniques is the name of methods and reactors which can easily be assembled from common laboratory objects, used for kinetic measurements from the 1850s on, and still used for slow reactions. Schematic diagram of a typical apparatus is shown in Fig. 7.2.

One of the reagents or pure solvent is filled in the reactor vessel; other reagents (often only one) are made to flow in from external container(s) in a pre-dosed quantity. The magnetic stirrer can homogenise the assembled mixture within 1–2 s. Storage of data in the computer is worth to start prior to complete mixing; starting time of the reaction can be determined based on the stored time-dependent data. The computer program receives digitally coded data from the instruments connected to the thermometer gauge and the concentration detector; reaction time is followed using the internal clock of the computer. Long before computer-controlled measurements, reaction time, detector signal and temperature were noted manually on a piece of paper. From the 1960s until about 2000, detector signal was recorded by an analogue device called *strip-chart recorder*; reaction time was coded by the constant ejecting velocity of the strip of paper. In case of looking

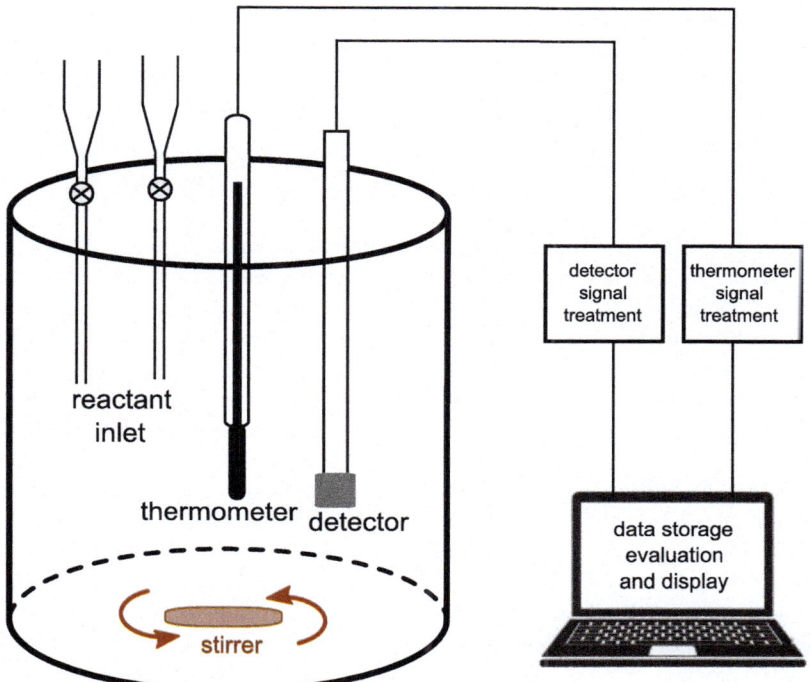

Fig. 7.2 Example for a classical batch reactor. Inflow of reactants is complete within a few seconds and the stirrer prepares a homogeneous mixture during inflow. The reactor is immersed into a thermostat (not shown); temperature is measured inside the reaction mixture. The detector can be an ion-selective electrode; a membrane electrode based on a redox reaction; a conductivity meter electrode cell; but also an in situ optical fibre sensor attached to a spectrophotometer. Signals from the thermometer and the concentration detector are first transformed into suitable digital signals than stored in the computer. The computer measures reaction time and displays the time-dependent signal in real time.

for data in old articles on kinetic experiments, it is well worth to check the method of data recording, especially for having an idea of the accuracy of reported data.

In classical measurements, the concentration detector should not always be immersed into the reaction mixture. Often, a method called *quenching* is used; when the reaction is stopped (or *quenched*) by a sudden change of conditions; e.g. by cooling part of the mixture withdrawn from the reactor, or changing its pH. The removed and quenched portion then can be used to determine concentration by a suitable method of titrimetry, a gas chromatograph or other type of chromatographs; or it can be introduced into an optical instrument. Especially in gas reactions, a common practice is to fill several small reactor tubes and quench them one by one at different reaction times to determine composition.

7.3.2 Flow Methods

Discharge flow method has been developed to study fast gas reactions, especially that of atoms or radicals. The reactor is a long tube: at one end, one of the reactants is introduced, while the other reactant is let in through a thinner tube (the so-called *injector*) somewhere into the middle portion of the reactor tube. A schematic diagram of the apparatus is shown in Fig. 7.3.

Reaction takes place from the entry point of the injector where reagents mix. Flow rates and pressure are adjusted so that mixed reagents move at a constant rate with a laminar flow (also called *plug flow*). As a consequence, reaction time is proportional to the distance between the point of mixing and the point of detection; it can be calculated from the distance x and the gas velocity v as $t = x/v$. (In Fig. 7.3, the distance x is marked by 'reaction distance'.) Thus, reaction time is easily controlled by changing the position of the injector (joined with an appropriate gasket to the reactor tube), while the detector can have a fixed position at the end of the reactor. An advantage of this arrangement is that – once the reaction time is fixed – detection time can be as long as it is necessary to achieve a desired signal-to-noise ratio.

To maintain laminar flow and provide thorough mixing of the reagents, gas pressure should be low enough; in the range of 100–1000 Pa. In a sufficiently narrow reactor, mixing is complete within a fraction of millisecond; thus, the usual time resolution of the discharge flow method is about 0.1–1 ms. This is the typical characteristic time of radical-molecule reactions; the method thus offers a convenient means to study them. To maintain constant gas flow rate and initial concentrations, high power pumps and precision flow control valves are needed. (Not shown in Fig. 7.3.)

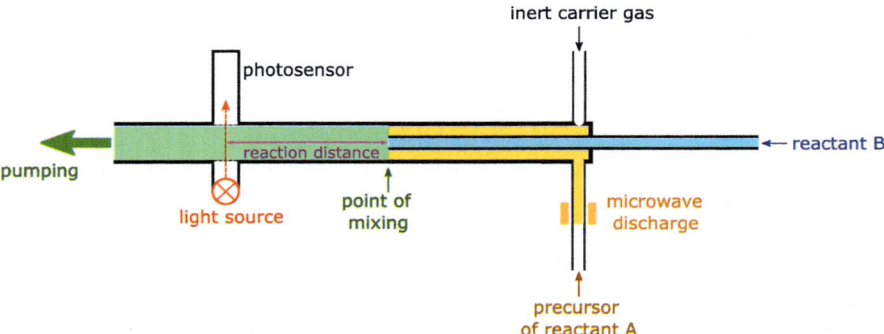

Fig. 7.3 Schematic diagram of a discharge flow reactor. Precursor of reactant A is introduced across a microwave discharge tube where atoms or radicals are formed prior to entering the reactor. Reactant B is introduced through a moving injector joined to the reactor tube by a convenient gasket, allowing to easily changing the position of mixing. Flow time between mixing and detection is the time of reaction from the beginning until concentration measurement. To control flow time and concentrations, gas inlet valves with flow controllers and a vacuum pump are applied. Concentration measurement can be realised also with other detection methods; photodetection is only shown as one possible example

7.3 Experimental Techniques and Devices

There is a need sometimes to conduct gas kinetic experiments at substantially higher pressures than optimal for fast mixing and laminar flow. Several researchers have developed reactors that can accommodate highly turbulent flow where mixing can be achieved at considerably high pressure. Evaluation of the measured results is more complicated in this case as consequences of the turbulent flow should also be taken into account.

To generate atoms from diatomic molecules or radicals from molecules with fully occupied valence shell, a convenient method is to let them flow across a microwave discharge tube. (See reactant A in Fig. 7.3.) The precursor of the reactant is mixed with helium as a carrier gas which largely facilitates dissociation or radical formation. To control the partial pressure of the reactant, additional carrier gas can be introduced after the microwave tube. This method is best to generate halogen or hydrogen atoms.

In the figure, a light source and a photodetector are shown to measure concentration. In a low-pressure gas, concentration of components is typically very low; thus, fluorescence detection is advantageous to have sufficient sensitivity. For this, the excitation is done by a light source (typically laser) emitting photons of appropriate energy, and a highly sensitive photodetector is needed. It is not seen in the figure, but in case of fluorescence detection, the detector is not in the way of excitation beam – as the high intensity exciting light would distort fluorescence detection – but at right angle to the exciting beam direction. Even in this case, there is usually a light absorbing black layer where the exciting beam hits the wall of the reactor.

Apart from fluorescence detection, many other methods can also be used to determine concentration; e.g. an EPR spectrometer discussed before, which is ideal to measure radical concentration. Another alternative is a mass spectrometer; it can be directly connected to the reactor tube (where pumping is indicated in the figure). Thus, it can do the necessary pumping, and the reaction mixture can directly enter the ion source of the mass spectrometer and detected in the mass analyser. As mentioned above, the mass spectrometer can not only measure concentration but also identify intermediate and product species. Discharge flow apparatus and common mass spectrometer are both relatively low-coast instruments, their use is also quite simple; thus they are widely used in the millisecond time domain to study gas reactions.

There exist also flow reactors to study liquid-phase reactions in solution. One of them – the so-called *continuous liquid flow reactor* – has the same principle as the discharge flow method for gas mixtures – but it is adapted to liquid solutions flowing in the tube, and no discharge is applied.

To make liquids flow, there is no need for pressure control nor pumping of the reaction products; it is enough to push the liquid using syringes with a constant velocity and let it freely flow across the tube. Syringe plungers can be moved using controlled velocity electric motors (similarly to the ones lifting car windows), or using regulated gas flow from a high-pressure gas cylinder. Reaction time – similarly to gas flow reactors – can be calculated from the distance between the point of mixing and the point of detection. If the mixing device is small enough, homogeneous mixing can be accomplished within a few microseconds; thus, time resolution is limited only by this and the flow rate. Using sufficiently narrow reactor tube, time resolution can be as low as a microsecond. Apart from photodetection, all kinds of detectors can be used which are capable of measuring low concentrations in very small volumes.

A major disadvantage of the continuous flow method is that its time resolution depends on the flow rate. In a reactor of sufficiently small-time resolution, high flow rate and a great throughput of the reaction mixture are needed. Thus, a large amount of reagents is also needed which is typically thrown away as waste (Fig. 7.4).

This disadvantage is avoided by the *stopped flow reactor*. The key difference with respect to the continuous flow reactor is that, after mixing, only a small reactor volume is filled with the reagents which then serves as a bulk reactor. Accordingly, the reaction takes place in a stationary, homogeneous phase.

Detector signal is sampled by a computer in *real time*; thus, time resolution is limited – apart from the sampling time – by the mixing time. The time between the start of the reaction at mixing and the standstill of the mixture in the reactor is called *dead time*. Dead time is not a limiting factor in case of first-order or pseudo-first-order reactions; in these cases, the rate coefficient can be calculated from the concentration–time profile even if the reaction has been progressed to some extant during mixing. (See introduction of Sect. 7.1). Dead time can be shortened by very efficient mixing and as small a reaction volume as possible. Most recent and most effective mixing devices are 0.1 mm diameter capillaries joined in a T shape (T-shaped micromixers). Reagents are injected through the two horizontal inlets of the T using 4–5 bar pressure; thus, the homogeneous mixture is prepared within 1 ms in the vertical capillary. Reactor volume can be reduced using a low-volume effluent syringe where the flow is stopped. A typical reactor has a volume of 5–20 µL;

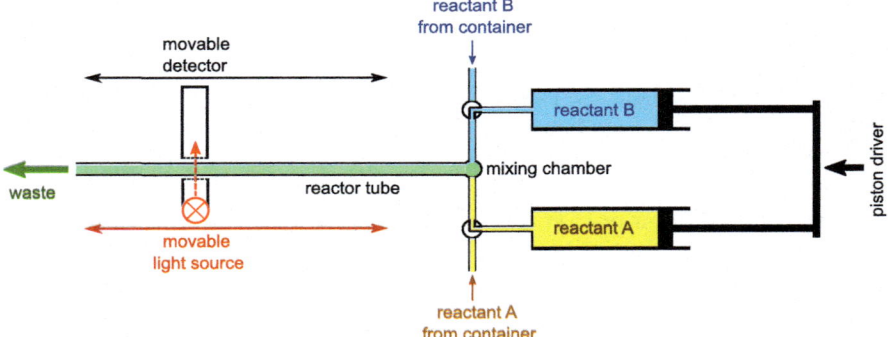

Fig. 7.4 Schematic diagram of a continuous liquid flow reactor. Reactants enter the tube from syringes and flow at a constant rate through it. Reaction time is coded in the distance between the mixing chamber and the detector and can be calculated from the flow rate. Light source and detector can be moved (together) along the reactor tube. (Detection can be made using other methods as well; photodetection shown is only one possibility.) Pistons are typically moved by an electric motor, but gas flow from a constant pressure bulb can also be used. Reactants can be refilled from large containers by turning the L-shaped valves

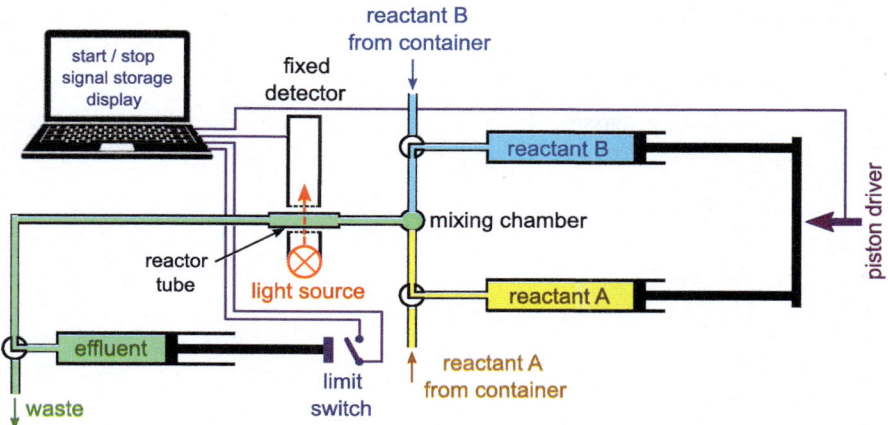

Fig. 7.5 Schematic diagram of a stopped flow reactor. Reactants enter the tube from syringes and quickly flow across to the effluent syringe. When the effluent syringe is filled, it activates the limit switch, stopping reactant feeding and starting signal recording to follow concentration changes. Light source and detector are fixed at the (small) reactor tube. Reaction time is measured by the internal clock of the computer. Detector signal related to concentrations is stored as a function of time. Detection can be continued for any predetermined time. Pistons are typically moved by an electric motor, but gas flow from a constant pressure bulb can also be used. Reactants can be refilled from large containers by turning the L-shaped valves and moving pistons backwards. At the same time, the effluent syringe is also emptied

together with all the tubing and the effluent syringe, not more than 50–100 µL mixture is needed to one fresh filling. Using the T-shaped micromixer and this low volume, dead time can be as low as 0.2 ms. Once the reactor is filled, observation time is not limited; reactions from a few milliseconds to several minutes can readily be followed (Fig. 7.5).

Once a time-dependent concentration measurement is completed, the plunger of effluent syringe is pushed back, thus the used reaction mixture leaves the syringe through an L-shaped bleeder valve, and it is ready to receive the next fresh filling. Syringes containing reactants can be refilled by turning their L-shaped valves. After returning the valves, a new filling and measurement can begin. The new filling also rinses remains of the previous filling from the reactor. Before data treatment, several experiments' data can be averaged to increase the signal-to-noise ratio and the precision of estimated kinetic parameters.

Stopped-flow apparatuses are popular and widespread; they are made by many manufacturers. Their most common application is enzyme-kinetic routine measurements (they have been developed for this purpose in 1964), but they are convenient to study redox reactions, complex formation and catalytic reactions as well. Apart from the typical absorbance or fluorescence spectrophotometers, almost all detectors mentioned in Sect. 7.1 can be combined with a stopped-flow apparatus – many of them contain built-in detectors.

7.3.3 Relaxation Methods

Section 7.2 contains a thorough discussion of temporal limit for mixing. According to conclusions of the discussion, kinetics of reactions shorter than 10–100 μs cannot be studied by mixing active reactants. Reactions taking place in the time domain from microseconds to femtoseconds can only be initiated avoiding mixing of the reagents. Typical solution for this problem is to pre-mix reactants so that at least one of them is in the mixture in a non-reactive form as a precursor. Thus, the reaction would not start at mixing but only after a very short interaction that activates the precursor. This fast initiation is discussed with the particular methods in subsequent sections.

The term *relaxation* is used for the kinetic method of rapid perturbation of an equilibrium mixture and subsequent study of the relaxation kinetics back to a new equilibrium state. To effectively and rapidly perturb equilibria, two major methods are used: *temperature jump* and *pressure jump*.

Temperature jump is usually achieved by an electric discharge across the liquid reaction mixture. To do this, a high-capacity (at least several μF) capacitor is charged to several thousand, or even several ten thousand volt potential and discharged through a small volume of the conducting reaction mixture. Depending on the volume, the discharge current can heat the mixture within 5–10 μs by 10 °C. After this perturbation, the reaction immediately proceeds towards a new equilibrium state at the elevated temperature. Different detectors can be applied to follow the temporal evolution in real time, which have the necessary time resolution. Temperature jump can also be attained by mixing in an inert component which is an excellent absorber of infrared radiation (e.g. SiF_4), and irradiating the mixture by a sufficiently high energy infrared laser pulse. SiF_4 molecules can very effectively and rapidly transfer their excess energy to other molecules, leading to a similar temperature jump as in the case of electric discharge. This method can also be used in case of non-conducting mixtures.

Pressure jump can be used to perturb equilibrium if the equilibrium depends on pressure; i.e. when there is a change in volume due to the reaction. Volume change in liquid phase is typically very small; if we want to change equilibrium concentrations, very large pressure change is needed in a very short time. First applications have been based on lowering pressure, as pressure can be diminished efficiently through a large cross-section valve or a rapidly broken diaphragm. In the twenty-first century, there are instruments where the piston of a very small volume cylinder can be operated by a piezoelectric actuator crystal. As it can be used to compress as well as expand cylinder volume, pressure jump can now be exercised in both increasing and decreasing directions. With a cylinder of 10–50 μL volume, up to 10,000 bar pressure change can be achieved both in increasing and decreasing directions. The time resolution can be fractions of milliseconds.

Though the *shock tube* is not operating using the principle of relaxation, its use is also based on a sudden increase of temperature and pressure. Shock tubes are used since 1900, but their chemical application only dates back to 1950. Schematic

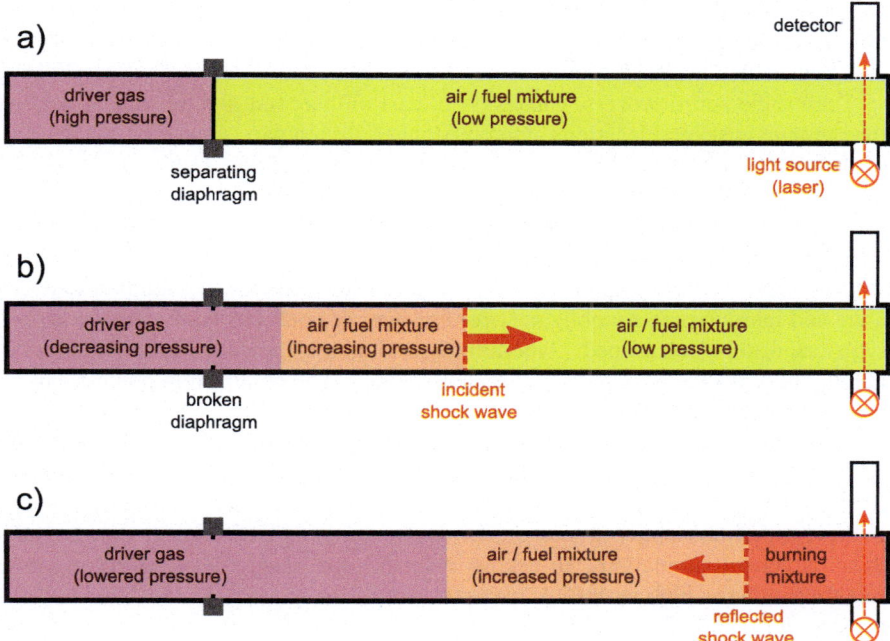

Fig. 7.6 Schematic diagram of a shock tube used to study combustion reactions. (**a**) High pressure compartment (containing driver gas) and low-pressure compartment (containing reaction mixture) of the tube divided by the diaphragm prior to bursting. (**b**) Following bursting of the diaphragm, the driver gas rapidly compresses the reaction mixture, heating it up. (**c**) The shock wave reflected from the end of the tube further heats the reaction mixture and ignites it. Pressure is continuously monitored at several locations along the tube; temperature is measured usually at the end. At the end of the tube, intermediate concentration is selectively measured (here: using a laser and photodetector)

diagram of the kinetic apparatus is shown in Fig. 7.6. Kinetic shock tubes are typically 5–10 m long, 50–100 mm diameter steel tubes. The high-pressure driving gas is in the smaller compartment of the tube; in case of kinetic experiments, it is typically high purity helium or argon. The longer, low pressure compartment containing the reaction mixture is separated by a diaphragm of limited pressure resistance – a thin metal foil or a ceramic plate.

To launch the pressure wave, either the driver gas pressure is increased until the foil bursts, or the ceramic plate is broken by an electromechanical pulse or a small detonation. As soon as the division is removed, the pressure wave compresses the driven gas with higher speed than that of the sound, and the traveling *shock wave* delivers a great part of its high energy to the reaction mixture. As a result, the pressure of the reaction mixture is increased and reactant molecules get translationally highly excited. This translational energy is rapidly distributed also among internal degrees of freedom (rotation and vibration), largely increasing temperature; sometimes up to a few thousand degrees centigrade. Pressure conditions can be adjusted – for example in case of ignition time measurements – so that the incident shock wave only pre-heats the air–fuel mixture; it is the reflected shock wave that ignites it.

During the entire process, gauges of at least millisecond time resolution follow the pressure at several locations along the longer, driven part of the tube. At the end of the tube, temperature is measured, and several intermediate (radical) concentrations are followed, typically by laser-induced fluorescence. Data evaluation is a computationally intensive task; differential equations modelling compression, propagation of the shock wave, change of temperature and the reaction mechanism should be numerically solved. In case of combustion experiments, ignition temperature and ignition time are obligatory results, but rate coefficients of several reaction steps can also be determined. Overall time resolution is typically in the millisecond range, but it can be extended also for shorter times. The combustion process can of course be followed for quite longer times after ignition.

With appropriate size scaling and the use of sensitive and energy-selective laser induced fluorescence methods, vibrational relaxation (time dependence of the transformation of translational to vibrational energy) can also be studied. Results obtained in these experiments have an important role in comparing kinetic measurements and their numerical simulations. In a kinetic shock tube, temperature of the reaction mixture can be as high as 4000 K; pressure can be increased from less than 1 bar in the driven part up to 10,000 bar.

7.3.4 Flash Photolysis and Laser Photolysis

Similarly to the temperature jump method, *flash photolysis* has also been developed after World War II. (Both techniques became feasible using high-capacity capacitors developed for long range radars during the war.) The schematic diagram of a flash photolysis apparatus is shown in Fig. 7.7.

At the time of their first use, there were no lasers available yet, thus a flash lamp was used for both excitation and detection – operating on the principle of high-current discharge across a low-pressure gas-filled tube; similarly to photoflash lamps. The capacitor was charged to a few hundred volt potential and discharged to provide the necessary current in the flash tube. High-intensity excitation flash reaches the reactor through a large surface to enhance efficiency of photolysis, while detection flash travels along the reactor tube to enhance absorption. The first detection method was to take a picture of the spectrum on a photographic plate or film in the spectrograph.

After sensitive enough spectrometers had been developed, the film was replaced by photodetectors, and their time-dependent signals were photographed on the cathode-ray tube display of an oscilloscope. There were still photographs used to evaluate signals: screenshots of the display. Dead time was limited by the temporal width of the exciting flash pulse, while time resolution by the temporal width of the

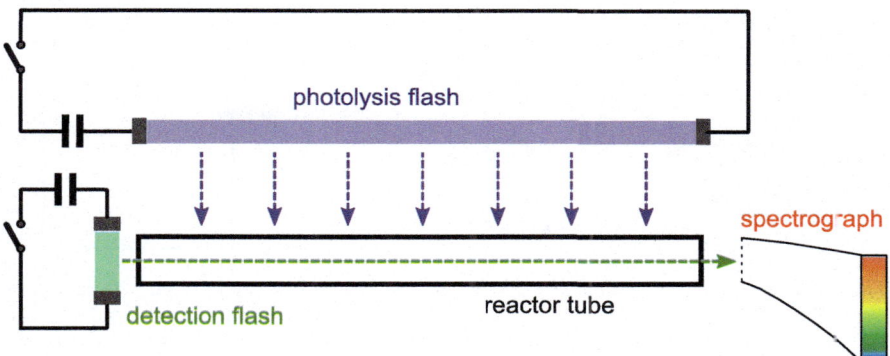

Fig. 7.7 Schematic diagram of a flash photolysis apparatus. Following the photolysis flash (and the simultaneous start of the reaction), a time delay circuit fires the detection flash lamp. This passes along the reactor tube and enters a spectrograph where the spectrum is recorded on a photographic plate or film. Spectral lines (or bands) are then analysed with a transmission densitometer on the developed photograph. Concentration of the components can be determined from the measured wavelength-dependent densities. Recording a series of spectra taken at different delay times provides concentrations as a function of time

detection flash. If both were optimised, a few microsecond time resolution was available.

Time resolution of kinetic measurements was revolutionised by the rapid evolution of pulsed lasers. Replacing excitation flash by high-intensity short-duration laser pulses, it is limited only by the pulse width and the precision of time measurement. At first, detection following *laser photolysis* was still realised by continuous white light and real-time detection. Since short-width pulsed lasers became relatively cheap, detection is also made using pulsed lasers; returning to time-delay between the excitation and detection pulses to control reaction time. The big difference is that the duration of a short laser pulse is much shorter in time than that of a flash lamp; thus, time resolution has also greatly improved. The schematic diagram of a laser photolysis apparatus is shown in Fig. 7.8.

Time resolution of the laser-photolysis apparatus depends on the duration of the excitation and detection pulses, but also on the available time resolution of the driver program on the computer, controlling the time delay between the two pulses. Development of mode-locked lasers in the 1960s made nanosecond time resolution available, which could be still controlled by the then available computer clock rate. (The highest clock rate by that time was around 1 GHz; enabling 1 ns time resolution – provided that a transient spectrum could be registered within one clock cycle.) Thus, mode-locked lasers were the first to open up the nanosecond time domain for kinetic measurements. Using very sophisticated detection methods, this resolution could be extended to a few tens of picoseconds, but shorter times cannot be measured at the beginning of the twenty-first century by computers – the fastest available clocks. (The fastest actual computer processors at this time are operating with no more than 10 GHz clock cycle.)

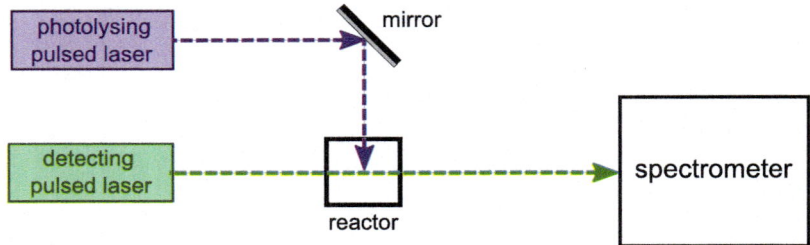

Fig. 7.8 Schematic diagram of a laser-photolysis apparatus. Following the photolysis laser pulse (and the simultaneous start of the reaction), the controlling computer fires the detection laser pulse with a predetermined delay. This travels along the reactor tube and enters a spectrometer; the digitised spectrum is then stored in the computer. Recording a series of spectra at different delay times provides concentrations as a function of time, which is readily calculated and displayed by the computer. Between subsequent measurements, the reactor can be slightly moved so that, at the crossing of the photolysis and detection beam, there is always a fresh (unreacted) reactant mixture

7.3.5 Kinetic Measurements at Femtosecond Timescale

Sub-picosecond time resolution in chemical kinetics was made possible by colliding pulse mode-locked (CPM) lasers, together with laser-pulse amplifying and subsequent pulse compression, available since the 1980s. With this development, kinetic experiments are possible within the entire 'chemical' timescale. Namely – as it is explained in the introductory part of this chapter discussing timescales – chemical reactions involve rearrangement of atomic nuclei, which takes at least 10 fs time.

Sub-picosecond duration laser pulses (also called *ultrashort* pulses) only reduce dead time and the time resolution of concentration detection; measuring and controlling time at the femtosecond scale should also be solved. (As explained above, 10 GHz clock cycle of the fastest computer processors make at highest only 0.1 ns resolution possible.) Femtosecond resolution can be achieved making use of the velocity of light; light travels 0.3 μm within 1 fs, which difference in distance can easily be controlled by precision micrometre screw delay lines or piezo-electric crystals. Thus, in a simple ultrafast kinetic apparatus, changing the path length of the detection pulse with respect to that of the exciting pulse by 0.3 μm changes the time difference between them by 1 fs and enables to control reaction time at this scale. Schematic diagram of this apparatus is shown in Fig. 7.9.

In case of ultrashort light pulses, the spectral width (energy range of photons propagating in the pulse) is also important. Laser pulses are shaped in time as waves; in optimal case, their shape is Gaussian. The shape in energy is also a wave: the Fourier transform of the temporal shape. Without going into mathematical details, it is only mentioned here that between the widths of the two waves, (Heisenberg's)

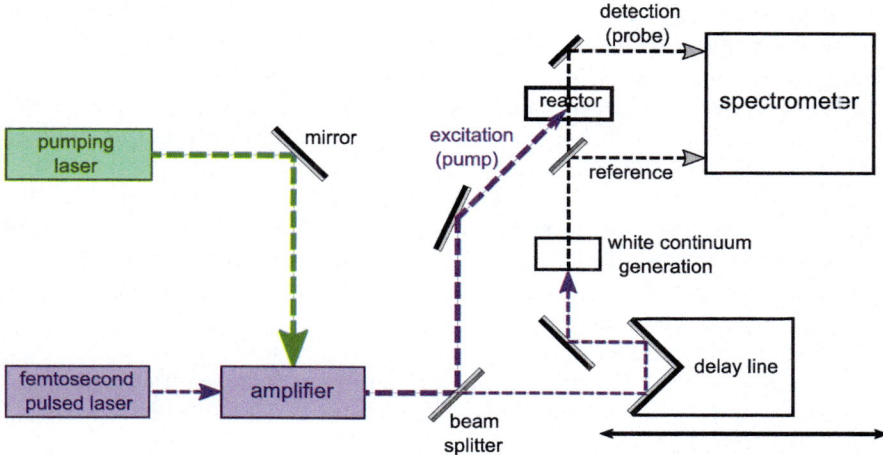

Fig. 7.9 Schematic diagram of a femtosecond resolution pump-and-probe experiment. Low-energy femtosecond laser pulses are amplified with the help of a pumping laser, and then re-compressed to the desired pulse width in the amplifier. A beam splitter directs the greater part of the photons into the reactor to excite the precursor and initiate reaction. The smaller part of photons passes through a delay line then enters a medium in which a white continuum is generated from the nearly monochromatic beam. This beam serves to record time-dependent spectra in the spectrometer. Part of this beam is directly introduced in the spectrometer and serves as reference to damp fluctuations in the detected spectra. Reaction time can be controlled by the delay line: 0.3 μm difference in the optical paths results in 1 fs time difference. Signals of many laser shots can be accumulated at a fixed delay, leading to an enhanced signal-to-noise ratio and more precise kinetic parameters

uncertainty principle holds.[5] According to this, the temporal width Δt and the width in energy $\Delta \nu$ (given in frequency units) should satisfy the inequality $\Delta t \, \Delta \nu \geq 1/(4\pi c)$. (In the formula, c is the velocity of light.) As an example, if a light pulse of $\Delta t = 40$ fs width at half maximum is needed, its spectral width (at half maximum) cannot be less than 1470 cm^{-1}. This means that, in case of 610 nm central wavelength, the spectral width of a 40 fs wide light pulse spreads from 585 to 641 nm wavelength. Thus, when exciting a precursor with this pulse, quite a few photons with wavelength between 585 and 641 nm enter the reaction mixture, not only those of 610 nm. If we should excite selectively – i.e. with well-defined photon energy – we should make a compromise by applying a temporally wider pulse. In the

---

[5]Werner Heisenberg (1901–1976) was a German physicist; one of the founders of quantum mechanics (Nobel-prize in physics 1932). He had oriented German nuclear research during WW II to building nuclear reactors instead of developing bombs. He published the *uncertainty principle* – later named after him – in 1927. This principle states the inherent property of wave mechanical quantities being pairs of Fourier transforms (like position and momentum, time and energy) that both of them cannot be concentrated in an arbitrarily small region. This has been later formulated by mathematicians as a theorem proved, extending it to any function pairs that are each other's Fourier transforms.

visible light domain, as a rule of thumb, for a 10 nm spectral width, minimum 100 fs duration light pulses should be used. Thus, if selective excitation is needed, 100 fs is a good compromise, but shorter than 10 fs pulses are no more applicable for selective excitation. In case the characteristic time of the studied process is shorter than this, special mathematical treatment – *deconvolution* – is needed to evaluate observed data to achieve a desired time resolution.

In a typical pump-and-probe apparatus of femtosecond time resolution, since the beginning of the twentieth century, diode-pumped solid-state lasers (typically of titanium-doped sapphire; $Ti:Al_2O_3$) are used. To amplify the ultrashort pulses generated, other solid-state lasers (e.g. neodymium-doped yttrium–aluminium garnet; Nd:YAG) are applied. To get two beams (for excitation and detection), beam splitters are used that transmit one part of the light and reflect the other part. For the spectral detection of concentrations, a quasi-white continuum beam is needed. This is generated by passing the very high energy, almost monochromatic pulse across some liquid (e.g. D_2O); the high photon density excites plenty of the molecular degrees of freedom which then emit practically within the pulse duration at almost all visible wavelengths. Another possibility is to use the nonlinear optical nature of the high photon density pulse to change the wavelength of the emerging pulse by an *optical parametric amplifier.*

The power of amplified high-energy ultrashort pulses – due to the very short duration – is in the TW/cm^2 (10^{12} W/cm^2) range. At this high power, pulse energy is fluctuating significantly; this is suppressed at detection using another beam splitter to direct a small part of detection light into the spectrometer without passing through the reactor. If the detected signal after having passed through the reactor is divided by this reference signal reflecting the actual laser intensity, fluctuations are greatly damped. By fixing a delay time (and thus the time of reaction until detection), laser pulses can be fired many times; thus the accumulated detected signals increase the signal-to-noise ratio until the desired accuracy of the measurements is achieved. Using solid-state lasers, the repetition frequency of laser shots can be as high as 1 kHz; enabling quite rapid collection of the repeated signals. To excite always a fresh, unreacted part of the reaction mixture, the reactor is slightly moved between laser shots. One possibility is to vibrate the reactor, but a more convenient way is to use a thin liquid jet instead of a reactor cuvette which is irradiated by the laser beam. In case of gas phase measurements, crossed molecular and laser beams are applied.

Sub-picosecond time resolution makes experimental observation of the course of elementary reactions possible. Comparing these experimental results with quantum mechanical modelling of the observed process, even potential energy surfaces can be determined; this also offers the possibility of observing transition states. For this reason, sub-picosecond kinetic measurements are of great importance. The science of experimental and theoretical treatment of femtosecond chemical processes also has a distinct name: *femtochemistry.*[6]

---

[6]The name femtochemistry was coined by the pioneer of the field, the 1999 Chemistry Nobel-prize winner, Egyptian-American physical chemist Ahmed Zewail (1946–2016), when trying to find a suitable name discussing with his colleague, Richard B. Bernstein (1923–1990), an American physical chemist.

Further Reading

1. Pilling MJ, Seakins PW (1995) Reaction kinetics. Oxford University Press, Oxford
2. de Paula J, Atkins PW (2014) Physical chemistry. 10th edn. Oxford University Press, Oxford
3. Silbey LJ, Alberty RA, Moungi GB (2004) Physical chemistry. 4th edn. Wiley, New York
4. Steinfeld JI, Francisco JS, Hase WL (1998) Chemical kinetics and dynamics. 2nd edn. Prentice Hall, Englewood Cliffs
5. Davis ME, Davis RJ (2003) Fundamentals of chemical reaction engineering. McGraw-Hill, New York
6. Andrews DL (1997) Lasers in chemistry. Springer, Berlin
7. Keszei E (2009) Efficient model-free deconvolution of measured femtosecond kinetic data using genetic algorithm. J Chemom 23:188–196

Index

A
Absorption spectrophotometers, 164, 165
Absorption spectrum, 164
Activation energies, 30–33, 36, 92, 93, 106, 117–119, 125, 134, 136, 142
Activation enthalpy, 31, 32, 135
Activation entropy, 31, 135
Activation Gibbs potentials, 134, 135
Activation process, 117–131
Activation volume, 32, 33
Active centres, 138, 139, 155
Active sites, 135, 137, 142, 147, 155, 156
Adsorption, 137, 139–141
Air–fuel mixture, 176
Alternator, 154
Ammonia synthesis, 155
Analytical solutions, 70, 76, 80, 87, 152
Anharmonic vibrations, 128
Arrhenius
  equations, 30–32, 36, 37, 130, 134
  plots, 30, 33
Attractive force, 10
Autocatalysis, 147–154
Autocatalyst, 147, 148, 150, 151
Autocorrelation, 67
Autoinhibition, 147, 148, 151
Autoinhibitor, 147, 148
Avogadro constant, 18, 19, 25, 34, 39, 49, 51

B
Batch reactors, 168, 169
Beam splitters, 179, 180
Beer, 133
Beer–Lambert law, 164

Berzelius, 133
Bimolecular reactions, 18, 33, 52, 117, 134
Binding pocket, 135, 142
Biocatalysts, 133, 135, 142
Biphasic reactions, 137
Bodenstein, 88, 101, 103, 105
Boltzmann constant, 8, 15, 25
Boltzmann distribution, 8, 24–28, 127
Boltzmann factor, 9
Born–Oppenheimer approximation, 24, 25, 29
Bouguer, 164
Branched chain reactions, 99, 105, 108
Branching, 71, 72, 75–77, 101, 106–108
  ratio, 75, 76, 106, 107
  reactions, 72, 76, 77
Bulk concentrations, 98, 139–141
Butterfly wings, 154

C
Calibration curve, 163
Canonical ensembles, 14, 17
Canonical partition functions, 15, 18
Carrier gas, 171
Cascade reactions, 72
Catalyst poisoning, 138
Catalysts, 6, 133–158
Catalytic constant, 147
Catalytic reactions, 133, 134, 136, 139, 147, 173
Cavitation, 119
CD spectrometer, 164
Chain carrier, 101, 104–106, 108
Chain length, 101
Chain reactions, 125, 133

Characteristic equation, 87, 109
Characteristic time, 161, 162, 170, 180
Chemical activation, 119
Chemical oscillations, 151, 153, 154
Chemical sink, 71
Chemical source, 71
Circular dichroism, 163, 164
Clock time, 162
Collinear reaction, 12–14, 20, 23, 24
Collision cross-section, 8
Collision frequency, 7–10, 30, 33, 34, 92
Collision theory, 7–10, 22, 23, 30, 52, 117, 122, 123
Combustion, 175, 176
Combustion engines, 154
Competitive reactions, 72
Composite reactions, 1, 2, 4, 41, 61, 62, 70–74, 90, 91, 94, 95, 98, 100, 106, 120, 124, 137, 148, 151, 154, 161
Conductivity meter, 165, 169
Confidence intervals, 66, 67, 69, 115, 157
Configurations, 10, 11, 16, 19, 20, 23, 24, 26, 29, 125
Configuration space, 26
Consecutive reactions, 72, 73, 77, 79, 80, 85
Continuous liquid flow reactor, 171, 172
Continuously-fed stirred tank reactor (CSTR), 154
Contour map, 13, 14, 19, 23, 26
Coupled system of differential-algebraic equations, 89
Curved Arrhenius plots, 33
Czakó, G., vii, 12

D

Dead time, 172, 173, 176, 178
Decadic molar absorptivity, 164
Deconvolution, 180
Definite integrals, 43, 96
Degenerate matrix, 110
Densitometer, 177
Density of states, 27, 29, 128
Desorption, 137, 139–141, 155
Detector signal, 163, 168, 172, 173
Detectors, 161, 162, 168–174
Determinant, 109, 110
Diatomic molecules, 10, 16, 102, 171
Differential method, 47, 66, 69
Diffusion, 136, 137, 140
Directed multigraph, 71, 100, 101
Discharge flow, 170, 171
Discharge tubes, 165

Dissipation, 154
Dissociative adsorption, 155
Distortion of the errors, 63
Docking, 135
Driver gas, 175
Durbin-Watson statistics, 67
Dynamical expression of the rate coefficient, 29

E

Eadie–Hofstee transform, 146
Eigenvalue, 86, 87, 108–112
Eigenvalue equation, 86, 87
Eigenvectors, 86, 87, 110–112
Elastic collisions, 8, 9
Electrocatalytic reactions, 137
Electron beam, 167
Electronic degeneracy, 17
Electronic partition function, 17
Electron paramagnetic resonance, 166
Electron spin resonance, 166
Elementary reactions, 1, 4–37, 39–41, 61, 71–73, 84, 92, 99, 161, 180
Emission, 163, 165
Emitted light, 164
Energetic molecules, 120, 123, 125, 127
Energy barrier, 8, 19, 20, 23, 26, 117
Enzyme, vi, 133, 135, 142–147, 156–158, 173
  catalysis, vi, 142–147
  regulation, 147
Enzyme-substrate complex, 143, 144
Equilibrium, 5, 6, 10, 11, 14–26, 28, 30, 31, 65, 73, 83, 84, 90–94, 98, 114, 135, 140–142, 155, 174
  constants, 5, 11, 14–19, 21, 30, 31, 65, 135, 140, 141, 155
  reactions, 5, 6, 10, 11, 14, 18–21, 23, 65, 135, 155, 174
Excess ratios, 63–65
Explicit solution, 42, 47, 49, 56, 60, 61, 76, 81, 113
Explosion limit, 107, 108
Explosions, 105–116
Extent of reaction, 3

F

Fall-off region, 121, 123
Fast pre-equilibrium approximation, 91–93
Feedback, 147, 148, 151, 153
Femtochemistry, 180
Femtosecond timescale, vi, 162, 178–180
Flash lamp, 176, 177

Index

Flash photolysis, 176, 177
Flooding, 62, 63, 148
Flow controllers, 170
Flow reactors, 168, 170–172
Fluorescence, 165, 171, 173, 176
   detection, 164, 171
   lifetime, 165
Fluorimeters, 165
Forbidden transition, 165
Formation of water, 2, 6, 106
Fourier transforms, 178, 179
Free energy, 5, 6, 14, 18, 134
Frequencies, 16, 21–23, 35, 118, 119, 124, 127–129, 166, 179, 180
Frequency factor, 9, 30, 32
Fructose, 156

G

Gas chromatograph, 169
General solutions, 42, 44, 48, 49, 55, 78, 81, 86, 87, 111, 149
Glucose, 156
Goniometer, 166
Graphical methods, 45, 47, 60–62

H

Half-life, 36, 44, 48, 50–53, 56, 57, 59
Hanes–Woolf transform, 146
Harmonic oscillator, 16, 21, 35
Heisenberg, 178
Heterogeneous catalysts, 135
Heterogeneous catalytic reactions, 47, 136, 137, 142, 145
High-energy radiation, 168
High-pressure limit, 122, 124, 129
Hinshelwood, 123, 124
Homogeneous catalysts, 135, 136
Homogeneous differential equation, 78
Hydrogen atoms, 10, 11, 13, 19, 155, 171
Hydrogen exchange reaction, 11, 24

I

Ignition time, 176
Impact parameter, 7
Implicit solutions, 44, 45, 50, 56, 60, 61, 103
Indefinite integrals, 44
Inelastic collision, 8
Inhibitors, 135, 136, 147

Initial concentrations, 41, 42, 44, 47, 50–54, 56, 57, 59, 60, 63–65, 69, 84, 96, 98, 113–115, 148–150, 156, 163, 170
Initial conditions, 42, 43, 47, 49, 55–58, 60, 61, 75, 77, 79–83, 85, 87, 108, 111–113, 149, 152
Initial value problem, 44, 96
Initiation, 1, 168, 174
Injector, 170
Integration constant, 42, 44, 49, 55, 81
Integration variable, 43, 44
Intermediates, 2, 4, 47, 133, 136, 143, 151, 155, 171, 175, 176
Internal vibrational-energy redistribution (IVR), 125, 127
International Union of Pure and Applied Chemistry (IUPAC), 2, 3
Invertase, 156
Ionising radiation, 118
Ion-selective electrode, 169
Isolation, 62, 63
Isothermal rate coefficient, 127
Isotope effect, 36

K

Kassel, 125, 126
Kinetic energy, 8, 9, 21, 27, 117, 123
Kinetic matrix, 86, 87, 109, 112
Kinetic theory of gases, 7, 30
Knocking, 154

L

Lambert, 164
Laminar flow, 170, 171
Langmuir-Hinshelwood equation, 141
Laplace transforms, 84, 85
Laser photolysis, 177, 178
Laser pulses, 162, 167, 168, 174, 177–180
Levenberg–Marquardt estimation method, 114
Libration, 118
Lindemann, 120–123, 130
Lindemann mechanism, 33, 121, 122, 124, 129, 143, 145
Lindemann-Hinshelwood mechanism, 124, 127
Linearization, vi
Lineweaver–Burk transform, 146
Liquid jet, 180
Logarithmic transformation, 47
Lotka, 152

Lotka-Volterra equations, 152
Low-pressure limit, 122
L-shaped valves, 172, 173

M
Magnetic stirrer, 167, 168
Marcus, 127, 128
Mass-action, vi, 40, 71, 98, 114, 148
Mass-action kinetics, 40, 71, 113, 148
Mass spectrometers, 166, 171
Maxwell distribution, 8
Mechanistic steps, 71, 73, 84, 113
Mechanochemistry, 119
Membrane electrode, 165, 169
Menten, 142, 144, 147, 156–158
Metastable equilibrium, 5
Michaelis, 145, 147
Michaelis-Menten mechanism, 142, 144, 147, 158, 159
Microcanonical rate coefficient, 127, 128
Micrometre screw, 178
Microwave discharge tube, 170, 171
Microwave radiation, 118
Mixing, 5, 6, 136, 157, 161, 162, 167, 168, 170–172, 174
Mixing of the reactants, 161
Mixing time, 65, 167, 172
Mode-locked lasers, 177, 178
Modified Arrhenius equation, 32, 36, 37
Molar concentrations, 3, 39, 48, 61, 142, 156, 157
Molecular beams, 166, 168
Molecular partition functions, 15–18, 21, 23
Molecularity, 40, 41
Monochromator, 164
Monolayer, 138

N
Nagy, T., vii, 35
Negative cooperativity, 147
Noise, 164
Non-homogeneous differential equation, 77, 80
Nonlinear chemical processes, 151, 154
Nonlinear estimation, 146
Non-linear parameter estimation, 63
Normal modes, 16
n-th order reaction, 40, 45, 47–50
Nuclear magnetic resonance (NMR), 166
Nucleon motion, 162
Nucleus–neutrino interactions, 162
Number densities, 8, 9, 33, 39

Number of collisions, 8, 9
Numerical derivation, 47
Numerical differentiation, 69
Numerical integration, 60, 61, 77, 96–99, 103, 114, 146, 152

O
Optical detection methods, 163
Optical fibre, 169
Optically active species, 156
Optical parametric amplifier, 180
Optical rotation, 156, 157, 163
Ordinary differential equations, 40, 42, 75, 97, 114
Oscillating reactions, 151
Oscillatory burning, 154
Oscilloscope, 176
Overall order, 62

P
Parallel reactions, 72
Parameter estimations, 47, 62, 63, 66, 69, 70, 114, 146
Partial fractions, 55, 60, 61, 149
Partial order, 62, 135
Particular solutions, 42, 44, 78, 81, 83, 86, 87, 111, 112, 149
Partition functions, 14–18, 21–23, 25, 27, 28, 32, 35
Pattern formation, 154
Periodic function, 151
Permutation with repetition, 126
Phase separation, 136, 137
Phase space, 26–28
Phosphorescence, 165
Photoactivation, 118
Photochemistry, 118
Photodetectors, 165, 171, 175, 176
Photodiodes, 165
Photodissociation, 128
Photographic plate, 176, 177
Photomultiplier, 165
Photoresistors, 165
Piezoelectric actuator, 174
Piezo-electric crystal, 178
Planck constant, 15
Plane-polarised light, 156, 163
Plug flow, 170
Polarimeter, 156, 163
Population dynamics, 151–153
Positive cooperativity, 147

Index   187

Potential energy, 10–13, 20, 23, 24, 27, 28, 180
Potential energy surface (PES), 10–14, 19, 20, 23–27, 29, 32
Precursor, 167, 170, 171, 174, 179
Pre-exponential factor, 30, 31
Pressure jump, 174
Product valley, 11, 12, 20, 26
Progress variable, 54, 59, 60
Propagation, 23, 146, 167, 176
Propagation of error, 67
Protein molecules, 135, 142
Pseudo-first-order reaction, 62
Pseudo-order, 148
Pulse radiolysis, 118
Pulsed lasers, 118, 165, 177
Pump-and-probe apparatus, 180

Q

Quantum mechanical modelling, 180
Quasi-equilibrium, 10, 11, 14, 18–26, 28
Quasistationaritätsprinzip, 88
Quasi-steady-state approximation (QSSA), 88, 144
Quenching, 169

R

Radiation chemistry, 118
Radicals, 136, 165, 166, 170, 171, 176
Radioactivity, 1, 118
Ramsperger, 125, 127
Rate-coefficient matrix, 86
Rate coefficients, 9, 22, 23, 25, 27–37, 39, 47–50, 52, 53, 57, 62, 63, 65–67, 69, 70, 90, 91, 98, 103–105, 113, 120–131, 139, 140, 144, 147, 150, 153, 161, 163, 168, 172, 176
Rate constant, 9
Rate-determining step, 93, 94, 122, 155
Rate equations, 39–62, 65, 67, 69, 70, 120–122, 138, 143, 148, 149, 151–153
Rate of appearance, 4
Rate of consumption, 4
Rate of conversion, 3
Rate of decomposition, 89, 92, 102
Rate of formation, 89, 92, 99, 102, 104, 120, 138, 143, 144
Rate of the reaction, 3, 9, 22, 106, 120, 128, 134, 144, 148, 159
Rational algebraic fraction, 55, 149
Reactant valley, 11, 12, 20, 23

Reaction mechanism, 1, 71, 76, 93, 95, 99, 101, 135, 148, 155, 176
Reaction paths, 11, 133, 134, 136, 142
Reaction rates, 4, 7–9, 22, 24–27, 36, 37, 39, 41, 47, 53, 58, 130, 136, 138, 140, 141, 145, 146, 155–159, 168
Reactive collisions, 8–10
Reactors, 161–163, 166–173, 176–180
Reduced mass, 8, 34, 123
Reflection, 163
Refractive index, 163
Refractometer, 163
Relaxation methods, 174–176
Repulsive force, 10
Residual errors, 67–69
Resonance fluorescence, 165
Reverse reactions, 73
Reversible reactions, 73, 83, 84, 88, 125
Rice, 125, 127
Rotational constants, 16, 35, 36
Rotational partition function, 15, 16
Rotational quantum number, 128
Rotational symmetry factor, 16
Round-off error, 97
RRK theory, 125, 127, 128
RRKM theory, 128
Runge–Kutta method, 97

S

Scattering angle, 166
Schrödinger equation, 29
Selective excitation, 180
Selectivity, 142, 165
Sensitivity, 165, 171
Serial reactions, 72
Shock tubes, 174–176
Shock wave, 175, 176
Signal-to-noise ratio, 170, 173, 179, 180
Skeleton model, 154
Solid-state lasers, 180
Sonication, 119
Sonochemistry, 119
Specific rotations, 156, 157
Spectrophotometers, 164, 169, 173
Spin angular momentum, 166
Splitting, 166
Stable equilibrium, 5
Standard deviations, 67, 98, 115, 146
Starting parameters, 114
Statistical thermodynamics, 10, 14, 20, 125
Steady-state concentration, 90, 102, 143

Steady-state hypothesis, 89, 90
Step-like function, 151
Stochastic kinetics, 98
Stoichiometric equations, 2–4, 18, 19, 39, 52, 53, 58–60, 71, 98, 99, 114, 134, 152
Stoichiometric numbers, 2–4, 39, 41, 48, 49, 53, 61
Stopped flow reactor, 172, 173
Strip-chart recorder, 168
Student-distributed variable, 67
Student's t-distribution, 66, 115
Sub-picosecond time resolution, 178, 180
Substrates, 142, 144–147, 156, 158, 159
Sucrose, 156
Sugar inversion, 163
Surface concentrations, 137–139, 163
Surface coverage, 138–140
Surface reactions, 137–140
Surface site, 137
Symbolic computation programs, 112
Symbolic computations, 85, 87
Syringes, 171–173
Systematic deviations, 67, 68
System of ordinary differential equations, 74, 98

T
Temperature jump, 174, 176
Temporal limit of mixing, 167
Termination, 133
Termolecular reactions, 40, 41, 92
Third body, 100, 107
Time delay circuit, 177
Time resolution, 161, 166, 168, 170, 172, 174, 176–178, 180
Time scales, 136, 178
Titrimetry, 169
Tolerance, 97, 98, 114
Transient, 177
Transition states, 11, 12, 14, 19–26, 28, 30, 32, 33, 35, 36, 39, 40, 71, 125–128, 180
Transition state theory (TST), 10–30, 32, 35, 36, 117, 162
Translational partition functions, 15, 21

Truncated standard molecular partition function, 22, 25
Truncation error, 97
T-shaped micromixers, 172
Tunnelling, 24
Turbulent flow, 171

U
Ultrashort pulses, 178, 180
Unbranched chain reactions, 99, 136
Uncertainties, 47, 63, 67, 69, 164
Uncertainty principle, 24, 178, 179
Unimolecular gas reactions, 33, 119, 122, 125, 129
Unimolecular reactions, 19, 32, 50, 72–74, 80, 117, 120, 123, 125, 127, 137, 141, 143, 145

V
Vacuum pump, 170
van't Hoff equation, 31
van't Hoff method, 47
Velocity of light, 16, 129, 178, 179
Velocity of sound, 167
Vibrational degrees of freedom, 16, 124
Volterra, 152

W
Wave number, 35
Wilhelmy, 163

X
X-rays, 167

Z
Zeeman effect, 166
Zero-order reaction, 47, 48
Zero-point energy, 35, 128
Zewail, 180

The manufacturer's authorised representative in the EU is Springer Nature Customer Service Centre GmbH, Europaplatz 3, 69115 Heidelberg, Germany. If you have any concerns regarding our products, please contact ProductSafety@springernature.com

Printed and bound by CPI Group (UK) Ltd, Croydon, CR0 4YY

25/03/2026

02078169-0014